Cándido Piñeiro

Dynamics of Entire Functions

Also of Interest

Linear Algebra
Vector and Inner Product Spaces
Saurabh Chandra Maury, 2024
ISBN 978-3-11-151570-0, e-ISBN (PDF) 978-3-11-151603-5

Modelling Stochastic Uncertainties
From Monte Carlo Simulations to Game Theory
Mohammed Elmusrati, 2024
ISBN 978-3-11-158470-6; e-ISBN 978-3-11-158505-5

Differential Geometry
Frenet Equations and Differentiable Maps
Muhittin E. Aydin and Svetlin G. Georgiev, 2024
ISBN: 978-3-11-150089-8; e-ISBN 978-3-11-150185-7

Abstract Algebra
With Applications to Galois Theory, Algebraic Geometry, Representation Theory and Cryptography
Gerhard Rosenberger, Annika Schürenberg and Leonard Wienke, 2024
ISBN: 978-3-11-113951-7; e-ISBN 978-3-11-114252-4

Hybrid Information Systems
Non-Linear Optimization Strategies with Artificial Intelligence
Edited by: Ramakant Bhardwaj, Pushan Kumar Dutta, Pethuru Raj, Abhishek Kumar, Kavita Saini, Alfonso González Briones and Mohammed K. A. Kaabar, 2024
ISBN: 978-3-11-132979-6; e-ISBN 978-3-11-133113-3

Cándido Piñeiro

Dynamics of Entire Functions

The Fractal Sets of Julia and Mandelbrot

DE GRUYTER

Author
Prof. Dr. Cándido Piñeiro
41927 Mairena del Aljarafe
Seville
Spain
candidopg52@gmail.com

All the figures in the book have been obtained by the author using MATLAB®

ISBN 978-3-11-168943-2
e-ISBN (PDF) 978-3-11-168968-5
e-ISBN (EPUB) 978-3-11-169029-2

Library of Congress Control Number: 2025930584

Bibliographic information published by the Deutsche Nationalbibliothek
The Deutsche Nationalbibliothek lists this publication in the Deutsche Nationalbibliografie; detailed bibliographic data are available on the Internet at http://dnb.dnb.de.

Cover image: Cándido Piñeiro
Typesetting: VTeX UAB, Lithuania

www.degruyter.com
Questions about General Product Safety Regulation:
productsafety@degruyterbrill.com

Dedicated to M.ª Ángeles and Ángela

Preface

In the theory of iteration of complex functions, we may consider two main aspects: the iteration of rational functions and that of transcendental entire functions. Although there are certain similarities, the dynamics of transcendental entire functions notably differ from the dynamics of rational functions, mainly due to the point at ∞. For rational functions, ∞ is dynamically the same as any other point. However, for entire transcendental functions, ∞ is an essential singularity. In fact, Picard's theorem tells us that every neighborhood of ∞ is mapped by a transcendental entire function over the whole plane except for at most one point, whereby this function takes any other value infinitely often. This behavior produces substantial differences in the dynamics and the geometry of the Julia set for these two classes of functions. Therefore, when writing a book on complex dynamics, we must choose either the first or the second option.

The iteration theory of entire functions originated in the paper by Fatou [41] and was developed mainly in the works of Baker [5, 6, 7, 8, 9, 10]. A major period of activity began in 1980 after Mandelbrot first used computer graphics to explore complex dynamics. Since then, research on this subject has undergone rapid development, and a vast number of significant papers have been published. Excellent introductions to iteration of rational functions have become available, such as those written by Beardon, Milnor, Carleson and Gamelin, and Steinmetz. Nevertheless, there are very few systematic books on the dynamics of entire functions, such as [63, 47]. Unfortunately, reading these books is often difficult for nonspecialists since their proofs are seldom clearly written and readers struggle to fully understand the arguments.

This book is a comprehensive introduction to the theory of iteration of entire functions and is intended to introduce the reader to the key topics in the field and to form a basis for further study. In general, the proofs are more detailed and, therefore, the book will also help mathematicians to become acquainted with complex dynamics. In no sense is this manuscript a complete account of the subject. Nevertheless, the book may also be useful to young researchers in this field before they tackle more specific works. The proofs of certain overly complicated theorems have been deliberately omitted. By assuming their results, we can swiftly advance towards interesting ideas in complex dynamics. In these cases, a suitable reference is given.

The reader is assumed to be familiar with the basic topics of the theory of complex variables as well as the basic notions regarding topology. Normal families of analytic functions, univalent functions on the unit disc, and the value distribution theory constitute essential tools for the study of the dynamics of entire functions. So that the book is self-contained, all the necessary background material regarding those subjects are studied in Chapters 2 and 3, and Appendix A, respectively.

In Appendix C, the necessary rudiments of Matlab RGB images are explained in order to create computer graphics of different sets considered in the book, such as the sets of Julia and Mandelbrot. In this Appendix, a gallery is also included where beautiful and

https://doi.org/10.1515/9783111689685-201

spectacular images are shown. All these images have been obtained by the author using Matlab, most of which are revealed here for the first time.

Finally, in Appendix D, we give hints to selected exercises.

At the end of each chapter, we include *Bibliography remarks* where we list the bibliography used in each section. Anyway, in these remarks, we omit those sections that deal with topics considered in most books.

The book is an expanded version of notes based on the series of lectures given by the author at the University of Huelva (Spain), in the period 2017–2020. In writing this work, extremely useful texts have been employed, mainly from books by Beardon [13], Calerson–Gamelin [20], Hua–Yang [47], Milnor [61], and Morosawa [63]. For further explanations, we recommend Hayman's book [42] for the value distribution theory, and Alexander's book [3] for historical information.

I am indebted to professor Luis Bernal for his help and suggestions to improve the proofs of some results. I am thankful to professor Fernando Muñoz who read the entire manuscript with care and insight. Finally, I am grateful to professor Enrique Serrano for many helpful discussions about typing the manuscript in LaTex.

Seville, Spain
February 2025

The author

Contents

List of Special Symbols

$AV(f)$	Asymptotic values of f, p. 115
$\mathcal{B}$	Class of bounded singular-type functions, p. 123
$\mathbb{B}(f)$	Filled Julia set of f, p. 77
$C(z_0, r)$	Circle of radius r with center z_0, p. 1
$CV(f)$	Critical values of f, p. 115
$d_c(z, w)$	Chordal distance, p. 9
$\mathbb{D}_r(a)$	Disc of radius $r > 0$ with center $a \in \mathbb{C}$, p. 1
$\mathbb{D}$	Unit disc, p. 1
$\mathbb{D}^*$	Punctured unit disc, p. 1
$\mathcal{E}$	Class of transcendental entire functions, p. 1
$\mathbb{F}(f)$	Fatou set of f, p. 77
$\mathbb{I}(f)$	Escaping set of f, p. 87
$\mathbb{J}(f)$	Julia set of f, p. 77
$\mathbb{L}(U)$	Set of limit functions, p. 104
$\mathbb{M}$	Mandelbrot set, p. 150
$M(r,f)$	Absolute maximum of $\lvert f\rvert$ in $\overline{\mathbb{D}}_r$, p. 1
$m(r,f)$	Proximity function, p. 26
$N(r,f)$	Counting function, p. 25
$\overline{N}(r,f)$	Counting function disregarding multiplicity, p. 34
$O(x)$	Quantity with absolute value $\leq C\lvert x\rvert$, p. 1
$O_f^+(z)$	Forward orbit of z, p. 2
$O_f^-(z)$	Backward orbit of z, p. 82
$\Omega(f, z^*)$	Attractive basin of z^*, p. 43
$\Omega_0(f, z^*)$	Immediate attractive basin of z^*, p. 43
$PV(f)$	Picard exceptional value of f, p. 79
$\mathbb{R}^-$	Negative real semiaxis, p. 135
$\mathcal{S}$	Class of finite singular type functions, p. 123
$S^+(f)$	Postsingular set of f, p. 118
$\operatorname{sing}(f^{-1})$	Set of singular values of f, p. 115
$T(r,f)$	Characteristic function, p. 26
$\lfloor x \rfloor$	Integer part function, p. 1

https://doi.org/10.1515/9783111689685-202

1 Introduction

In this chapter, we develop the basic notions of iteration theory in the general framework of metric spaces, including the notions of conjugacy and completely invariant set. Next, we present all the necessary background material about the Riemann sphere and Möbius transformations. Finally, we do a brief historical overview of complex dynamics.

1.1 Terminology and notations

If (X, d) is a metric space, given $x_0 \in X$ and $r > 0$, we denote by $B(x_0, r)$ the *open ball* of radius r with center x_0 defined by

$$B(x_0, r) = \{x \in X : d(x, x_0) < r\}.$$

If A is a subset of X, the *boundary* and the *complement* of A are denoted by $\partial(A)$ and A^c, respectively. The *diameter* of a nonempty set $A \subset X$ is

$$d(A) = \sup\{d(x, y) : x, y \in A\}.$$

The real line and the complex plane are denoted, respectively, by $\mathbb{R}$ and $\mathbb{C}$. The *extended plane* is denoted by $\hat{\mathbb{C}} = \mathbb{C} \cup \{\infty\}$. Further, $\mathbb{D}_r(a)$ is the disc (open) of radius $r > 0$ with center $a \in \mathbb{C}$ (sometimes $\mathbb{D}(a, r)$ is more convenient). When the center is the origin, we shall shorten to $\mathbb{D}_r$. The *unit disc* is denoted by $\mathbb{D}$, the *punctured unit disc* $\mathbb{D} \setminus \{0\}$ by $\mathbb{D}^*$, and, finally, $C(z_0, r)$ denotes the circle of radius r with center z_0.

A connected and open set $\Omega \subset \mathbb{C}$ is called a *region or domain*.

By $O(x)$ we denote a function of x whose absolute value is bounded by $C\,|x|$, with C being a suitable positive constant, while the notation $o(f)$ means that $o(f)/f \to 0$. For any real number x, $\lfloor x \rfloor$ is the largest integer that is smaller than x.

As usual, a function $f : \mathbb{C} \to \mathbb{C}$ which is analytic at every point $z \in \mathbb{C}$ is called *entire*, while a *meromorphic function* in a region Ω means a quotient of two analytic functions in Ω. An entire function (meromorphic) is called *transcendental* if it is not polynomial (rational). We will denote the class of transcendental entire functions by $\mathcal{E}$. For every entire function f and $r > 0$, we denote by $M(r, f)$ the absolute maximum of $|f|$ in $\overline{\mathbb{D}}_r$, namely $M(r, f) = \max\{|f(z)| : |z| \le r\}$.

A map is called *univalent* (or *conformal*) if it is analytic and one-to-one.

1.2 Iteration of a function

Let X be a metric space and let $f : X \to X$ be a continuous function in X. For every integer $n \ge 0$, we define f^n inductively by setting: (i) $f^0(x) = x$ and (ii) $f^n(x) = f(f^{n-1}(x))$ for all $x \in X$ and $n \in \mathbb{N}$ (notice that $f^1 = f$); $(f^n)_n$ is called the sequence of *iterates* of f.

https://doi.org/10.1515/9783111689685-001

If we choose a point $x_0 \in X$ and apply repeatedly f, we obtain a sequence of points

$$x_0, x_1 = f(x_0), \quad x_2 = f(x_1), \ldots,$$

or, expressing x_n in terms of x_0, $x_n = f^n(x_0)$. The sequence $(x_n)_n$ is called the *(forward) orbit* of x_0 under f and is denoted by $O_f^+(x_0)$. About this sequence we can ask the following questions:

- If an orbit (x_n) converges to z^*, what can we say about the point z^*?
- Given two initial points $x_0, y_0 \in X$, consider the sequences $x_n = f^n(x_0)$ and $y_n = f^n(y_0)$. If x_0 and y_0 are close enough, do the orbits $(x_n)_n$ and $(y_n)_n$ have a similar behavior for $n \to \infty$?

Among others, these are the problems that are considered in iteration theory.

With respect to the first question, it is easy to show that the limit of a convergent orbit is a very special point for f. Suppose that the orbit (x_n) of some point x_0 converges to $x^* \in X$. By the continuity of f, it follows that

$$x^* = \lim_{n\to\infty} x_n = \lim_{n\to\infty} f(x_{n-1}) = f\Big(\lim_{n\to\infty} x_{n-1}\Big) = f(x^*).$$

Then $f(x^*) = x^*$ and we say that x^* is a *fixed point* of f. Obviously, the orbit of x^* is a constant sequence.

The following examples show that the behavior of the orbits, if we take the initial point x_0 close to a fixed point x^*, may be very different depending on the choice of the fixed point x^*.

Examples 1.2.1. (1) Let $X = \mathbb{D}_{1/2}(0)$ and $f(z) = z^2/(z^2+1)$ for $z \in X$. First, we show that $f(X) \subset X$. Given $z \in X$ with $z \neq 0$, we have

$$\frac{|f(z)|}{|z|} = \frac{|z|}{|1+z^2|} \le \frac{|z|}{1-|z|^2} = g(|z|),$$

where $g(x) = x/(1-x^2)$ for $x \in (-1,1)$. As $g'(x) = (1+x^2)/(1-x^2)^2$, it follows that g is strictly nondecreasing in $(-1,1)$ and, therefore,

$$\frac{|f(z)|}{|z|} \le g(|z|) < g(1/2) = \frac{2}{3}.$$

This proves the inequality

$$|f(z)| \le (2/3)|z| \tag{1.1}$$

for all $z \in X$, and we conclude that $f(z) \in X$.

By induction on n, we can obtain

$$|f^n(z)| \le (2/3)^n |z|, \tag{1.2}$$

which implies that $f^n(z) \to 0$ for all $z \in X$. Finally, note that $f(0) = 0$.

(2) Let $X = \mathbb{C}$ and $f(z) = z^2$. In this case, f has two fixed points $z_1^* = 0$ and $z_2^* = 1$. The iterates of f can be obtained easily

$$f^2(z) = f(f(z)) = f(z^2) = z^4.$$

In general, $f^n(z) = z^{2^n}$. Thus we have
(a) If $|z| < 1$, then $\lim_{n\to\infty} f^n(z) = 0$.
(b) If $|z| > 1$, then $\lim_{n\to\infty} f^n(z) = \infty$.

As in the above example, there exists a neighborhood of the fixed point $z_1^* = 0$ (the unit disc) such that all the orbits starting in $\mathbb{D}$ converge to the fixed point $z_1^* = 0$. Nevertheless, the behavior of the orbits starting close to the fixed point $z_2^* = 1$ is rather different. If D is a small disc with center 1, the orbit of all $z \in D$ escapes from D. Indeed, given $z \in D$, if $|z| \neq 1$, by (a) and (b) the orbit of z escapes from D; while if $|z| = 1$, we can take $\theta \in \mathbb{R}$ such that $|\theta|$ is small and $z = \exp(\theta i)$. It is obvious that there exists an $n \in \mathbb{N}$ such that

$$f^n(e^{\theta i}) = e^{2^n \theta i} \notin D.$$

(3) Let $X = \mathbb{C}$ and $f(z) = z + z^2$. Notice that $z^* = 0$ is the unique fixed point of f. We will show that there exist orbits starting close to the origin with a completely different behavior. Consider a small disc $\mathbb{D}_r$. If $x \in (-r, 0)$, then $x < f(x) < 0$ and, iterating the argument, we obtain $x < \cdots < f^{n-1}(x) < f^n(x) < 0$. Hence the orbit $(f^n(x))$ is convergent and, as we have seen above, its limit must be the unique fixed point of f. This proves that every point of the segment joining $-r$ to 0 is attracted by z^*. On the other hand, if $x > 0$, we have $f(x) = x^2 + x > x$ and again we may prove that $(f^n(x))$ is increasing and necessarily divergent since there is only a finite fixed point. Consequently, every orbit starting in the segment joining 0 to r escapes from $\mathbb{D}_r$.

In Example 1.2.1 (1) and (2), it is obvious that the convergence of the orbits to the fixed point is uniform on a certain disc. In view of those examples, the following definitions arise in a natural way.

Definition 1.2.2. Let X be a metric space, $f : X \to X$ continuous, and x^* a fixed point of f.
(a) We say that x^* is attracting if there exists $B(x^*, r)$ so that $f^n(x) \to x^*$ uniformly on $B(x^*, r)$.
(b) We call x^* repelling if there exists $r > 0$ such that, for each $x_0 \in B(x^*, r)$ $(x_0 \neq x^*)$, there is an $n \in \mathbb{N}$ such that $f^n(x_0) \notin B(x^*, r)$.

In the case of complex functions $f : \mathbb{C} \to \mathbb{C}$, we will see that a fixed point z^* is of one or the other type depending on $|f'(z^*)|$. Usually, the complex number $f'(z^*)$ is called the *multiplier* of the fixed point z^*.

In general, an orbit may be non-convergent. It suffices to imagine two points x_0 and x_1 such that $x_1 = f(x_0)$ and $x_0 = f(x_1)$. Then the orbit of x_0 has the form

$$x_0, x_1, x_0, x_1, \ldots,$$

that is, it is a periodic orbit with period 2.

Example 1.2.3. If $f(z) = iz^3$, take $z_0 = 1$ and $z_1 = i$. Notice that $f(1) = i$ and $f(i) = 1$. Then the points $z_0 = 1$ and $z_1 = i$ are periodic with period 2. The orbit of $z_0 = 1$ is

$$1, i, 1, i, \ldots$$

Definition 1.2.4. Let $f : X \to X$, $x_0 \in X$, and $p \geq 2$ a natural number such that $f^p(x_0) = x_0$. We say that x_0 is a periodic point and, if p is the minimum positive integer satisfying $f^p(x_0) = x_0$, we call p the period of x_0. If this is the case, the orbit of x_0 is the successive repetition of the finite sequence

$$\{x_0, f(x_0), \ldots, f^{p-1}(x_0)\}.$$

The set $\{x_0, f(x_0), \ldots, f^{p-1}(x_0)\}$ is called a p-cycle of f (fixed points are 1-cycles).

If $m > n \geq 1$ and $f^m(x_0) = f^n(x_0)$, the point x_0 is called preperiodic. The orbit of x_0 has the form

$$x_0, \ldots, f^{n-1}(x_0), f^n(x_0), \ldots, f^{m-1}(x_0), f^n(x_0), \ldots, f^{m-1}(x_0), \ldots,$$

that is, from the nth entry the orbit consists of the successive repetition of the finite sequence $f^n(x_0), f^{n+1}(x_0), \ldots, f^{m-1}(x_0)$.

Examples 1.2.5. (1) If $f(z) = e^{2\pi i/7}z^2$, $z_0 = 1$ is a periodic point with period 3:

$$f(1) = e^{2\pi i/7}, \quad f(e^{2\pi i/7}) = e^{6\pi i/7}, \quad f(e^{6\pi i/7}) = e^{14\pi i/7} = e^{2\pi i} = 1.$$

The orbit of $z_0 = 1$ is

$$1, e^{2\pi i/7}, e^{6\pi i/7}, 1, e^{2\pi i/7}, e^{6\pi i/7}, \ldots$$

(2) Let $f(z) = \pi i e^z$. The orbit of the origin is

$$0, \pi i, -\pi i, -\pi i, -\pi i, \ldots,$$

and $z_0 = 0$ is preperiodic.

In later chapters, we will study the behavior of the orbits starting close to a periodic point and see the significant role that periodic points play in the dynamics of a function.

Now we turn to the second question asked at the beginning of this section. Given a function $f : X \to X$, consider the subset of X defined by

$$\mathbb{B}(f) = \{x \in X : (f^n(x))_n \text{ is bounded}\}.$$

The answer to that question is negative for the set $\partial(\mathbb{B}(f))$. Indeed, if x_0 is a boundary point of $\mathbb{B}(f)$, there exist points $x_1 \in \mathbb{B}(f)$ and $x_2 \notin \mathbb{B}(f)$ as close to x_0 as one wants. So, there is a point, x_1 or x_2, close to x_0 so that its orbit has a rather different behavior as $n \to \infty$. In Chapter 6, we will see that every complex entire function f divides the plane in two complementary subsets which are completely invariant: the stable set (or Fatou set) $\mathbb{F}(f)$ and the Julia set $\mathbb{J}(f)$. In the latter set, the answer to the mentioned question is negative. Indeed, the behavior of f is chaotic in $J(f)$.

We finish this section by considering a general property of fixed points that we will need later. Suppose that X is an arbitrary set and $f : X \to X$ a continuous map. We denote by $\mathrm{Fix}(f)$ the set of all fixed points of f. Given another function $g : X \to X$, we say that f and g are *permutable* if $f \circ g = g \circ f$.

Theorem 1.2.6. *If $f, g : X \to X$ are permutable functions, then*

$$f(\mathrm{Fix}(g)) \subset \mathrm{Fix}(g).$$

Proof. Let $x \in X$ be a fixed point of g, then

$$g(f(x)) = f(g(x)) = f(x). \qquad \square$$

If g is the limit of a sequence of iterates of f, it is easy to prove that f and g are permutable.

1.3 Conjugation

The concept of conjugation is fundamental in the study of dynamics systems. Indeed, from a qualitative, or topological, point of view, the only dynamical properties that are interesting to be studied are those that are invariant by conjugation. In complex dynamics, often the difficulty of a problem is reduced through a suitable conjugation. We will use this idea often throughout the book.

Definition 1.3.1. Let X and Y be metric spaces and $f : X \to X$ and $g : Y \to Y$ continuous maps. We say that f is conjugate to g ($f \sim g$) if there exists a homeomorphism $h : X \to Y$ such that $h \circ f = g \circ h$.

It is obvious that $\sim$ is an equivalence relation. Note that if $h \circ f = g \circ h$, then $f^n = h^{-1} \circ g^n \circ h$ for every $n \in \mathbb{N}$. The importance of this lies in the fact that two functions belonging to the same equivalence class have the same dynamics. As an example, next we will prove that the type of a fixed point (attracting or repelling) is a conjugation invariant property.

1) If $x_0 \in X$ is a fixed point of f, then $y_0 = h(x_0) \in Y$ is fixed for g.
   Indeed, $g(y_0) = g(h(x_0)) = h(f(x_0)) = h(x_0) = y_0$.
2) If x_0 is an attracting fixed point of f, then $y_0 = h(x_0)$ is attracting for g.

As x_0 is attracting, there exists an $r > 0$ such that $\lim_{n\to\infty} f^n(x) = x_0$ uniformly on $B(x_0, r)$. By the continuity of h^{-1}, there is an $s > 0$ such that

$$y \in B(y_0, s) \quad \Longrightarrow \quad h^{-1}(y) \in B(x_0, r).$$

Then $f^n(h^{-1}(y))$ converges uniformly on $B(y_0, s)$ to x_0 since $h^{-1}(y) \in B(x_0, r)$. Finally, as h is continuous, we have

$$\lim_{n\to\infty} g^n(y) = \lim_{n\to\infty} h \circ f^n \circ h^{-1}(y) = h\Big(\lim_{n\to\infty} f^n(h^{-1}(y))\Big) = h(x_0) = y_0,$$

uniformly on $B(y_0, s)$.

3) If x_0 is a repelling fixed point for f, such is also $y_0 = h(x_0)$ for g.
By hypothesis, there is an $r > 0$ with the property that, for each $x \in B(x_0, r)$ $(x \neq x_0)$, there exists an $n \in \mathbb{N}$ such that $f^n(x) \notin B(x_0, r)$. By continuity, $h(B(x_0, r))$ is a neighborhood of y_0. Therefore, there exists an $s > 0$ such that

$$B(y_0, s) \subset h(B(x_0, r)). \tag{1.3}$$

Given $y \neq y_0$ in $B(y_0, s)$, we can choose an $x \in B(x_0, r)$ $(x \neq x_0)$ such that $y = h(x)$. As x_0 is repelling, there is an $n \in \mathbb{N}$ for which $f^n(x) = f^n(h^{-1}(y)) \notin B(x_0, r)$. Since $f^n(h^{-1}(y)) = h^{-1}(g^n(y))$, we conclude that $h^{-1}(g^n(y)) \notin B(x_0, r)$ and, therefore, $g^n(y) \notin h(B(x_0, r))$. This and (1.3) imply that $g^n(y) \notin B(y_0, s)$.

When we are considering complex functions, there is another notion of conjugacy of interest that is called conformal conjugacy.

Definition 1.3.2. Let U and V be two regions of the complex plane, $f : U \to U$ and $g : V \to V$ analytic functions. We say that f is conformally conjugate to g if and only if there exists a univalent function h from U onto V such that $h \circ f = g \circ h$.

By the chain rule, if f is conformally conjugate to g, then the respective fixed points have the same multiplier, too.

1.4 Completely invariant sets

Let X be an arbitrary set and let $f : X \to X$ be a map. We will say that $E \subset X$ is *completely invariant* if E and X/E are invariant, that is, $f(E) \subset E$ and $f^{-1}(E) \subset E$ hold. The next proposition lists several basic properties of completely invariant sets.

Proposition 1.4.1. *Under the above conditions, the following statements hold:*

(i) *If f is onto, then every completely invariant set E satisfies $E = f(E) = f^{-1}(E)$.*
(ii) *If $\phi : Y \to Y$ is bijective and $g = \phi \circ f \circ \phi^{-1}$ is conjugate to f, then ϕ maps completely invariant sets for f to completely invariant sets for g.*
(iii) *The union and intersection of an arbitrary family of completely invariant sets of X are completely invariant.*

Proof. For arbitrary f and E, we have $f(f^{-1}(E)) \subset E$. The reverse inclusion follows easily since f is onto. Now notice that

$$E = f(f^{-1}(E)) \subset f(E). \qquad \square$$

Given a subset E of X, property (iii) allows us to define the notion of the smallest completely invariant set containing E as the intersection of all completely invariant subsets of X containing E. This set is called the *completely invariant set generated by* E and we denote it by $\langle E\rangle$. We will need the following obvious property:
(M) If $E \subset F$, then $\langle E\rangle \subset \langle F\rangle$.

To facilitate the study of completely invariant sets, we are going to introduce the following binary relation: for any $x, y \in X$, $x \sim y$ means that there exist nonnegative integers m and n such that $f^m(x) = f^n(y)$. It is evident that $\sim$ is an equivalence relation and, for each $x \in X$, we denote by $[x]$ the equivalence class of x (some authors call $[x]$ the *grand orbit* of x).

Proposition 1.4.2. *Let X be an arbitrary set and let $f : X \to X$ be a map. The following statements hold:*
(i) *The class $[x]$ is completely invariant.*
(ii) *The class $[x]$ is the smallest completely invariant set containing x.*

Proof. (i) For each $y \in X$, we have $y \sim f(y)$, thus $x \sim y$ and $x \sim f(y)$ are equivalent. In terms of sets, this means that

$$y \in [x] \quad \Longleftrightarrow \quad f(y) \in [x].$$

From this, we deduce that $y \in [x]$ if and only if $y \in f^{-1}([x])$. Thus we have proved the equality $[x] = f^{-1}([x])$. On the other hand, the inclusion $f([x]) \subset [x]$ is obvious and, in consequence, $[x]$ is completely invariant.

(ii) If $y \in [x]$, there exist m and n such that $f^m(y) = f^n(x)$. Then

$$y \in f^{-m}(f^n(\{x\})) \subset f^{-m}(f^n(\langle x\rangle)) \subset \langle x\rangle,$$

where we have used the complete invariance of $\langle x\rangle$ in the last step. Since $[x]$ is completely invariant, the minimality of $\langle x\rangle$ yields $\langle x\rangle \subset [x]$. $\square$

The first consequence of the above proposition is that a set $E \subset X$ is completely invariant if and only if E is a union of classes, $E = \bigcup_{x\in E}[x]$. In fact, from property (M) it follows that $\bigcup_{x\in E}[x] \subset \langle E\rangle$ and, by the minimality property, the reverse inclusion is obvious.

Proposition 1.4.3. *Let X be a metric space and let $f : X \to X$ be an open continuous map. If $E \subset X$ is completely invariant, then so are $\mathring{E}$, $\partial(E)$, and $\overline{E}$.*

Proof. Let E be completely invariant. By continuity, $f^{-1}(\mathring{E})$ is open in X. On the other hand, $f^{-1}(\mathring{E}) \subset f^{-1}(E) \subset E$, which implies that $f^{-1}(\mathring{E}) \subset \mathring{E}$. Since $f(\mathring{E}) \subset f(E) \subset E$ and f is open, this argument can be repeated with the open set (in X) $f(\mathring{E})$. Thus $\mathring{E}$ is completely invariant.

As $\overline{E} = E \cup \partial(E)$, it suffices to prove that $\partial(E)$ is completely invariant. To this end, let $x \in \partial(E)$ and let V be a neighborhood of $f(x)$. We choose a neighborhood U of x such that $f(U) \subset V$. There exist $x_i \in U$ $(i = 1, 2)$ such that $x_1 \in E$ and $x_2 \notin E$. Then $f(x_1)$ belongs to $V \cap E$ since E is completely invariant. Now notice that $f(x_2)$ belongs to V but it is not in E, thus $\partial(E)$ is invariant. Finally, using that f is open, by the same argument we may prove that $f^{-1}(\partial(E)) \subset \partial(E)$. □

Lemma 1.4.4. *If f is surjective and the grand orbit of x_0 is finite, then $[x_0]$ is a cycle.*

Proof. If $[x_0]$ is finite, since it contains the orbit of every point $x \in [x_0]$, it follows that this orbit must be finite. Then $[x_0]$ only contains cycles. In particular, x_0 is periodic with period n. If $x_0, x_1, \ldots, x_{n-1}$ is the corresponding n-cycle, we will prove that $[x_0] = \{x_0, x_1, \ldots, x_{n-1}\}$. In fact, if $y_0 \in [x_0]/\{x_0, x_1, \ldots, x_{n-1}\}$, y_0 is also periodic and, if m denotes the period of y_0, all the cycle $\{y_0, f(y_0), \ldots, f^{m-1}(y_0)\}$ is contained in $[x_0]$. As $y_0 \sim x_0$, there exist positive integers k and h such that $f^k(y_0) = f^h(x_0)$. Thus there is a $j \leq n-1$ such that $x_j = f^k(y_0)$, which necessarily implies that y_0 belongs to the cycle $\{x_0, x_1, \ldots, x_{n-1}\}$. This shows that the grand orbit is a cycle. □

Example 1.4.5. Let $f(z) = (z - a)^n + a$ with $n \geq 2$. Then $z = a$ is a fixed point of f and the set formed by all preimages of a reduces to the point a. Thus $z = a$ is a point whose grand orbit is the singleton $\{a\}$.

1.5 The Riemann sphere

Every polynomial $P(z)$ of degree $n \geq 2$ has $n+1$ fixed points: the n solutions of the equation $P(z) = z$ (counting its multiplicity) and the point ∞ (as we will prove in Chapter 6). That is why in the study of the dynamics of rational functions it is necessary to include the point ∞. Then it is natural to consider the extended complex plane $\hat{\mathbb{C}} = \mathbb{C} \cup \{\infty\}$. From a topological point of view, we only need to define the neighborhoods of ∞. We will say that a set $U \subset \hat{\mathbb{C}}$ is a neighborhood of ∞ if it contains ∞ and the set $\{z \in \mathbb{C} : |z| > r\}$ for some $r > 0$.

The following simple example shows vividly that the point ∞ has no special significance considering the extended plane $\hat{\mathbb{C}} = \mathbb{C} \cup \{\infty\}$.

Example 1.5.1. Let $f(z) = z^2$. We have seen that $f^n(z) \to 0$ if $|z| < 1$, and $f^n(z) \to \infty$ if $|z| > 1$. If we set $f(\infty) = \infty$, then ∞ and 0 are attracting fixed points of f. Now consider the map $T(z) = 1/z$ and note that T maps 0 to ∞, and vice versa. On the other hand, T is one-to-one, analytic, and satisfies $T \circ f \circ T^{-1} = f$. That is, T conjugates f to itself. This shows that the behavior of f close to 0 or ∞ is the same.

In relation with the notion of compactness, we will need to use the fact that the topology of $\hat{\mathbb{C}}$ is metrizable. To this end, we introduce the Riemann sphere and define the chordal metric. Besides, in this way we get a geometric image of $\hat{\mathbb{C}}$.

Consider the unit spherical surface S^2 of equation $x_1^2 + x_2^2 + x_3^2 = 1$. The plane $x_3 = 0$ will be the complex z-plane. The point $N(0,0,1)$ of the sphere is called the north pole. Given a point $P(x_1, x_2, x_3)$ of the unit sphere, the line joining N to P will pierce the plane $x_3 = 0$ at the point P^*, which corresponds to the complex number $z = x + yi$ (see Figure 1.1). The map $z \to P$ is called the *stereographic projection* of $\mathbb{C}$ into S^2. If we put in correspondence N with ∞, this produces a bijective map from the extended complex plane onto the Riemann sphere S^2.

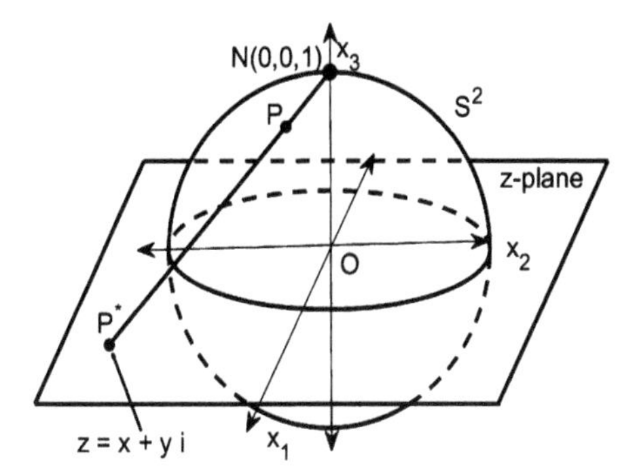

Figure 1.1: The stereographic projection of $\hat{\mathbb{C}}$ onto S^2.

Elementary computations yields

$$x_1 = \frac{2\,\mathbb{R}e(z)}{1+|z|^2}, \quad x_2 = \frac{2\,\mathbb{I}m(z)}{1+|z|^2}, \quad x_3 = \frac{|z|^2-1}{1+|z|^2}.$$

Now we proceed to define a metric on the extended plane. Given two complex numbers $z, w \in \mathbb{C}$, the *chordal distance* $d_c(z,w)$ is equal to the Euclidean distance between its stereographic projections on the Riemann sphere. An easy exercise in vector geometry yields

$$d_c(z,w) = \frac{2|z-w|}{\sqrt{(1+|z|^2)(1+|w|^2)}}$$

and

$$d_c(z,\infty) = \lim_{w\to\infty} d_c(z,w) = \frac{2}{\sqrt{1+|z|^2}}.$$

When we use the chordal distance $d_c(\cdot,\cdot)$, ∞ loses any special meaning. This statement is reinforced by noting that the map $h : z \in \hat{\mathbb{C}} \to 1/z \in \hat{\mathbb{C}}$ preserves the chordal distance (see Exercise 7).

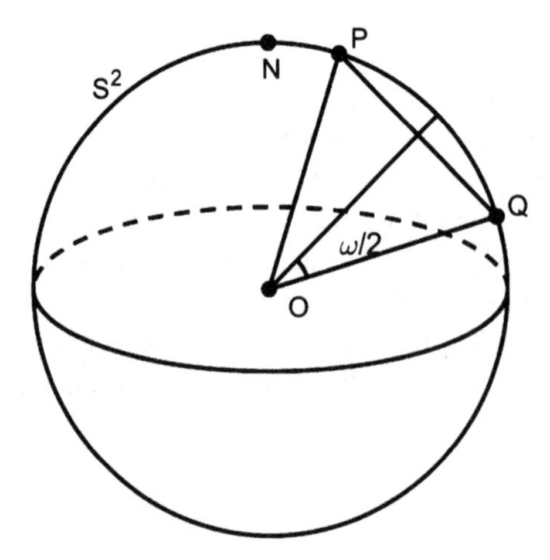

Figure 1.2: The spherical metric.

There is an alternative metric on $\hat{\mathbb{C}}$, the *spherical metric*, that is equivalent to the chordal metric. By definition, $d_0(z,w)$, if $z, w \in \hat{\mathbb{C}}$, is the shortest distance between P and Q in the unit sphere (an arc of a great circle), where P and Q are the stereographic projections of z and w on the Riemann sphere (see Figure 1.2). Suppose that the chord joining P to Q subtends an angle ω at the origin, then we have

$$d_0(z,w) = \omega \quad \text{and} \quad d_c(z,w) = 2\sin(\omega/2).$$

Therefore, we have the following relationship between both metrics:

$$d_c(z,w) = 2\sin\left(\frac{d_0(z,w)}{2}\right).$$

The next inequalities are more useful

$$\left(\frac{2}{\pi}\right)d_0(z,w) \le d_c(z,w) \le d_0(z,w). \tag{1.4}$$

The proof of these inequalities is Exercise 1.

1.6 Möbius transformations

If P and Q are polynomials with complex coefficients, the function defined by $f(z) = P(z)/Q(z)$ for all $z \in \mathbb{C}$ is called a rational function (P and Q are not simultaneously the zero polynomial). We will suppose that P and Q have no common zeros. We put $f(\infty) = \lim_{z\to\infty} f(z)$, and $f(z_0) = \infty$ if z_0 is a zero of $Q(z)$. In this way, $f(z)$ is a function defined from $\hat{\mathbb{C}}$ into itself. Therefore, the extended complex plane is the adequate framework to study the dynamics of these functions.

The number $d = \max\{\deg(P), \deg(Q)\}$ is called the *degree* of f. Usually, the rational maps of degree 1 are called *Möbius transformations*:

$$z \to T(z) = \frac{az + b}{cz + d}, \quad ad - bc \neq 0.$$

Recall that we set $T(\infty) = a/c$ and $T(-d/c) = \infty$. Then T is a bijective and bicontinuous map from the extended plane $\hat{\mathbb{C}}$ onto itself. Next we list some basic properties of Möbius transformations:

1) They map lines and circles of the complex plane to lines or circles.
2) The set of all Möbius transformations is a group (with respect to the composition of transformations).
3) The Möbius transformations of the form

$$T(z) = \frac{az - \bar{c}}{cz + \bar{a}}, \quad |a|^2 + |c|^2 = 1,$$

are isometries in $(\hat{\mathbb{C}}, d_c)$ (Exercise 7).

Throughout the book, automorphisms of the unit disc $\mathbb{D}$ and the upper half-plane $\mathbb{H} = \{z = x + yi : y > 0\}$ play an important role. Now we will see that these automorphisms are special Möbius transformations.

Theorem 1.6.1. (1) *Automorphisms of* $\mathbb{D}$ *are the following Möbius transformations:*

$$w = T(z) = e^{i\theta}\left(\frac{z - a}{1 - \bar{a}z}\right),$$

where $a \in D$ *and* $\theta \in [0, 2\pi]$.

(2) *Automorphisms of* $\mathbb{H}$ *are the conformal automorphisms of the upper half-plane having the form*

$$w = T(z) = \frac{az + b}{cz + d}, \quad a, b, c, d \in \mathbb{R} \text{ and } ad - bc > 0.$$

(3) *The half-plane* $\mathbb{H}$ *is conformally isomorphic to the unit disc* $\mathbb{D}$ *through the map*

$$\phi : w \in \mathbb{H} \to z = \frac{i - w}{i + w} \in \mathbb{D},$$

whose inverse is

$$\phi^{-1} : z \in \mathbb{D} \to w = i(1 - z)/(1 + z) \in \mathbb{H}.$$

Proof. (3) If z and $w = u + iv$ are related in this way, we have

$$|z| < 1 \Leftrightarrow |i - w|^2 < |i + w| \Leftrightarrow u^2 + (1 - v)^2 < u^2 + (1 + v)^2 \Leftrightarrow v > 0.$$ □

The iterates of a Möbius transformations T can be obtained explicitly and, in consequence, we can see easily that for most of these functions the orbits $(T^n(z_0))_n$ are convergent for every initial point z_0.

Example 1.6.2. Let $R(z) = (3z-2)/(2z-1)$. By induction, we can obtain the explicit form of the iterates of R, namely

$$R^n(z) = \frac{(2n+1)z - 2n}{2nz - (2n-1)} = 1 + \frac{z-1}{2nz-(2n-1)}.$$

We see that $R^n(z) \to 1$ as $n \to \infty$, for all $z \in \mathbb{C}$.

The behavior of the function in the above example is not a coincidence. Indeed, there is a general result which we pass to study.

We will need the following Jacobi's classical result. First, we recall that the unit circle S^1 and $\mathbb{R}/2\pi\mathbb{Z}$ are isomorphic, where the latter set is endowed with the metric given by

$$d(\alpha,\beta) = |\alpha - \beta| \mod 2\pi\mathbb{Z} \quad \text{for all } \alpha,\beta \in \mathbb{R}/2\pi\mathbb{Z}.$$

The map $h : \alpha \in \mathbb{R}/2\pi\mathbb{Z} \to z = e^{\alpha i} \in S^1$ is a homeomorphism.

Let $\lambda \in \mathbb{R}$ and $T_\lambda(z) = z\exp(2\pi\lambda i)$ for $z \in S^1$. Then T_λ behaves quite different depending on the rationality or irrationality of λ. If $\lambda = p/q$, with p and q integers, then $T_\lambda^q(z) = z$. Thus every point is fixed by T_λ^q. However, in the irrational case, we have the following result.

Theorem 1.6.3 (Jacobi's theorem). *If λ is irrational, then every orbit under T_λ is dense in S^1.*

Proof. Given $z \in S^1$, note that the equality $T_\lambda^n(z) = T_\lambda^m(z)$ implies $(n-m)\lambda \in \mathbb{Z}$ and this yields $n = m$. Then the points of the orbit of z are pairwise different. Since S^1 is compact, every infinite subset must have a limit point. Thus, given $\varepsilon > 0$, there exist integers m and n for which $|T_\lambda^n(z) - T_\lambda^m(z)| < \varepsilon$. If we put $k = n - m$ (in case $n > m$), it follows from the above inequality that

$$|T_\lambda^k(z) - z| < \varepsilon. \tag{1.5}$$

On the other hand, notice that T_λ preserves distances in S^1. Consequently, (1.5) implies $|T_\lambda^k(z) - T_\lambda^{2k}(z)| < \varepsilon$. Hence the sequence of points $(T_\lambda^{nk}(z))_n$ produces a partition of the unit circle into arcs whose end points are at a distance less than ε. This concludes the proof since ε is arbitrary. □

A Möbius transformation R can have a unique fixed point (double) or two different fixed points. We consider each case separately.

1) **R has only one fixed point**

   (a1) If ∞ is the unique fixed point of R, then $R(z) = z + b$ (we assume that $b \neq 0$) and elementary calculus shows that $R^n(z) = z + nb$. Then $R^n(z) \to \infty$ for all z.

(b1) If $R(z)$ has a unique fixed point z^* which is finite, then the Möbius transformation $g(z) = 1/(z - z^*)$ maps z^* to ∞. If we define $S(z) = g \circ R \circ g^{-1}(z)$, then ∞ is the unique fixed point of S and, therefore, we are in the case (a1). Thus $S(z) = z + b$ $(b \neq 0)$ and $\lim_{n\to\infty} S^n(z) = \infty$ for all z. On the other hand, since $S^n = g \circ R^n \circ g^{-1}$, we have

$$\begin{aligned}\lim_{n\to\infty} R^n(z) &= \lim_{n\to\infty} g^{-1} \circ S^n \circ g(z)\\ &= g^{-1}\Big(\lim_{n\to\infty} S^n(g(z))\Big) = g^{-1}(\infty) = z^*\end{aligned}$$

for all z.

Furthermore, the above argument gives us a method to get an explicit form of $R^n(z)$, $R^n(z) = g^{-1} \circ S^n \circ g(z)$.

2) **R has two different fixed points**

(a2) If the fixed points of R are 0 and ∞, then $R(z) = kz$ and $R^n(z) = k^n z$. For $z \neq 0, \infty$, we have

$$\begin{cases} R^n(z) \to 0 & \text{if } |k| < 1;\\ |R^n(z)| = |z| & \text{if } |k| = 1;\\ R^n(z) \to \infty & \text{if } |k| > 1.\end{cases}$$

In the case $|k| = 1$, there are two possibilities:

- k is an nth root of unity. Then $R^n(z) \equiv z$; or
- k is not an nth root of unity for all $n \in \mathbb{N}$. Now the points $R^n(z)$ are dense in the circle of radius $|z|$ with center at the origin (recall Jacobi's theorem).

(b2) If R has two different finite fixed points z_1^* and z_2^*, then consider the Möbius transformation given by $g(z) = (z - z_1^*)/(z - z_2^*)$. If $S = g \circ R \circ g^{-1}$, then S fixes 0 and ∞. As in the above case, the orbits $(R^n(z))$ are in one of the following situations:

- They converge to one of the fixed points of R.
- They are finite orbits.
- The points $R^n(z)$ form a dense subset of a circle (again by Theorem 1.6.3).

A Möbius transformation is called *elliptic* if it is conjugate to $z \to \lambda z$ with $|\lambda| = 1$ $(\lambda \neq 1)$. So, in view of the above study, nonelliptic Möbius transformations have the property that every orbit converges to some point $a \in \hat{\mathbb{C}}$ (it is easy to prove that the convergence is uniform on compact subsets of $\hat{\mathbb{C}}$).

We finish this section recalling the notion of a *derivative at the point* ∞. Suppose f is a rational function such that $f(\infty) = \infty$. If $T(w) = 1/w$, the function $F = T^{-1} \circ f \circ T$ is rational, satisfies $F(0) = 0$, and, therefore, it is analytic at the origin. By definition, $F'(0)$ is the derivative of f at the point ∞. If we use, in the about definition, another Möbius transformation mapping 0 to ∞, the value of $f'(\infty)$ does not change (see Exercise 12).

1.7 Historical overview

At the end of the nineteenth century and the beginning of the twentieth, the mathematicians Leau, Koenigs, Böttcher, among others, studied the dynamics of an analytic complex function in a neighborhood of a fixed point. Before the year 1906, little was known about global behavior, only a few easy examples were published by E. Schröder (1841–1902) and A. Cayley (1821–1895) in relation with the Newton method to obtain approximate values of the zeros of a polynomial [21, 22, 71, 72]. In the literature, it is usual to quote only Cayley as the predecessor of Julia and Fatou. Nevertheless, Cayley conceived the Newton method as a discrete process, while Schröder's works are more in line with the iteration of functions. In 1906, P. Fatou presented an interesting example, $f(z) = z^2/(z^2+2)$. He proved that, for almost all z, the iterations $f^n(z)$ are convergent to 0. However, for z belonging to a subset of $\mathbb{C}$ with zero measure, the orbit remains distant from 0. After the publication of this example, the interest in the study of complex dynamics increased.

In 1915, the Academy of Sciences of Paris announced a "Grand Prix des Sciences Mathématiques" that would be granted at the beginning of 1918 on the topic: *Iteration in one or several variables from a global point of view*. The Academy received three separate entries from Julia, Lattes, and Pincherle. Julia was the winner with his well-known work [51]: *"Memoire sur l'iteration des fonctions rationelles"*. The fundamentals of the global theory are due to Julia and Fatou. Fatou's extensive work "*Sur les equations fonctionelles*" was published in 1919 and 1920 [40]. A dispute between Fatou and Julia involved controversy over whose work took priority.

Between Fatou's and Julia's results and the explosion of research that started in around 1980, there is a 60-year gap. In this period, the most significant works are those of Siegel and Baker. In the 1940s, Siegel proved that Siegel discs really exist [74]. In the 1960s, Baker extended part of the work of Fatou and Julia to other classes of functions, namely entire and meromorphic.

In the late 1970s, the second period of major activity in complex dynamics started with Mandelbrot's work and, especially, with Douady and Hubbard's work (1982). One of the first images created through complex dynamics is the famous and spectacular image obtained by Mandelbrot, which can be observed in Figure 10.1. Although several of the results established in Mandelbrot's first work were inaccurate, he has the merit of being the first person to reveal the great complexity of the associated geometry with the parameter space of the quadratic family. His greatest achievement has been to show to a wide audience that fractals play a major role in numerous branches of mathematics. Clearly, the beautiful and intricate fractal images obtained with the aid of computers have greatly boosted interest in complex dynamics research. It could be stated that the floodgates were opened when computer graphics became available. In 1982, Douady and Hubbard introduced the name of the Mandelbrot set and established a solid foundation for its study. They proved that the Mandelbrot set and its complement are connected [34].

Although the study of the iteration of entire functions was initiated by Fatou [41] in 1926, it is in this second period when this study has become generalized. In this period, other authors deserve mention in addition to those already named: Sullivan, Bergweiler, Lyubich, and Eremenko, among others. In 1976, Baker proved the existence of wandering domains in the framework of transcendental functions [9], and, in 1985, Sullivan demonstrated that this is not possible in the case of rational functions [76].

We finish this brief history with a list of mathematicians that have contributed to the development of complex dynamics. As founders of complex dynamics we list Schröder, Poincaré, Koenigs, Leau, Lattes, Julia, Fatou, and Cremer. Among the more recent authors, we mention Siegel, Baker, Eremenko, Bergweiler, Lyubich, Hubbard, Sullivan, Herman, Douady, Arnold, and Yoccoz.

1.8 Exercises

1. Prove the following relation between the chordal and spherical metrics:
$$\frac{2}{\pi}d_0(z,w) \le d_c(z,w) \le d_0(z,w).$$
2. If $h(z) = z + 1/z$, prove that h is a conformal isomorphism from $A = \{z : |z| > 1\}$ onto $B = \mathbb{C} \setminus [-2,2]$.
3. Consider the function $P(z) = a_2z^2 + 2a_1z + a_0$ with $a_2 \neq 0$. Show that $P(z)$ is topologically conjugate to the quadratic map $Q_c(z) = z^2 + c$ through the conjugacy map $\phi(z) = a_2z + a_1$, where c is a suitable constant.
4. Determine the inverse of the stereographic projection.
5. We can define a stereographic projection in this other way. Consider the sphere S^2 with radius 1/2 and center $(0,0,1/2)$ that is tangent to the complex plane $\mathbb{C}$ at the origin. The point $N(0,0,1)$ will be referred to as the north pole of S^2. A correspondence T from S^2 onto $\hat{\mathbb{C}}$ is defined as follows: $T(N) = \infty$ and $T(P) = x + iy$ for every $P \in S^2$ $(P \neq N)$, with $(x,y,0)$ being the point that the line joining N to P intersects in the plane $z = 0$. Calculate the expressions of T and its inverse T^{-1}.
6. Let f be a bijection of a set X onto itself. Show that
$$[x] = \{f^n(x) : n \in \mathbb{Z}\},$$
for every $x \in X$.
7. Consider the Möbius transformations of the form
$$T(z) = \frac{az - \overline{c}}{cz + \overline{a}}, \quad |a|^2 + |c|^2 = 1.$$
Prove that: (1) the class of all these transformations is a subgroup of the group of all Möbius transformations and (2) they are isometries if we consider the chordal distance in $\hat{\mathbb{C}}$.
8. Let $f : \overline{\mathbb{D}} \to \overline{\mathbb{D}}$ be continuous on $\overline{\mathbb{D}}$ and analytic on $\mathbb{D}$. Show that:

(a) The equation $f(z) - Az = 0$ has only one solution in $\mathbb{D}$ for every $A > 1$.
(b) The function f has a fixed point in $\overline{\mathbb{D}}$.

9. Let $f(z) = z - z^2$. Prove that $\mathbb{D}_r(r)$ is invariant under f for every $r \in (0, 1/2]$.
10. Let $Q_c(z) = z^2 + c$ with $c \in (-2, 0)$. If β denotes the positive fixed point of Q_c, prove that the set $U = \{z : |z| \geq \beta\}$ is invariant under Q_c.
11. Let f be analytic in a convex region $\Omega \subset \mathbb{C}$ such that $|f'(z) - 1| < 1$ in Ω. Prove that f is univalent in Ω.
12. Let f be a rational function satisfying $f(\infty) = \infty$. (i) Prove that the value of $f'(\infty)$ does not depend on the Möbius transformation used in the definition. (ii) Calculate $f'(\infty)$.

Bibliography remarks

Sections 1.4, 1.6 are inspired by [13, Sections 1.2, 3.2]. The proof of Jacobi's theorem is based on [28, Section 1.3].

In Section 1.7, we follow [3].

2 Normal families

2.1 Normal families and equicontinuity

The notion of a normal family of functions is fundamental in the theory of complex iteration. Given a function $f : \mathbb{C} \to \mathbb{C}$, Fatou defined the stable set of f as the set formed by all points z_0 such that $(f^n)_n$ is normal in a neighborhood of z_0 and he proved that the complement of this set is the closure of the set of all repelling periodic points of f (for rational functions of degree equal to or greater than two). Later, those sets were called Fatou and Julia set, respectively.

The relationship of complementarity between the sets $\mathbb{F}(f)$ and $\mathbb{J}(f)$ is also valid for entire functions. This was proved later by Baker in 1968. Recall the second question we asked in the first chapter. Given two initial points z_0 and w_0, consider their orbits $z_n = f^n(z_0)$ and $w_n = f^n(w_0)$. If z_0 and w_0 are sufficiently close, do the sequences $(z_n)_n$ and $(w_n)_n$ have basically the same behavior as $n \to \infty$? The answer to this question seems to lead us to the notion of equicontinuity.

Definition 2.1.1. Given $f : \mathbb{C} \to \mathbb{C}$ and a disc $\mathbb{D}(z_0)$ with center z_0, we say that the sequence of iterates $(f^n)_n$ is equicontinuous in $\mathbb{D}(z_0)$ if, for every $\epsilon > 0$, there exists $\delta > 0$ such that

$$\left.\begin{array}{r} z, w \in \mathbb{D}(z_0) \\ |z - w| < \delta \end{array}\right\} \implies |f^n(w) - f^n(z)| < \epsilon \quad \text{for all } n \in \mathbb{N}.$$

If (f^n) is equicontinuous in $\mathbb{D}(z_0)$, the answer to the above question is positive in a neighborhood of z_0.

However, when the orbit of z_0 is divergent (or unbounded), equicontinuity is a very restrictive condition. Indeed, the equicontinuity of (f^n) implies that if w_0 is close to z_0, $f^n(w_0)$ is close to $f^n(z_0)$ for every $n \in \mathbb{N}$. But in the case $f^n(z_0) \to \infty$, it seems reasonable to say that the orbit $(f^n(w_0))$ has a similar behavior when it is also divergent. That is why, in the definition of equicontinuity, we will use the chordal metric (z and w may be far away in the complex plane and, nevertheless, their chordal distance is small).

Definition 2.1.2. Under the above conditions, we say that the sequence of iterates $(f^n)_n$ is d_c-equicontinuous in $\mathbb{D}(z_0)$ if, for every $\epsilon > 0$, there exists $\delta > 0$ such that

$$\left.\begin{array}{r} z, w \in \mathbb{D}(z_0) \\ |z - w| < \delta \end{array}\right\} \implies d_c(f^n(w), f^n(z)) < \epsilon \quad \text{for all } n \in \mathbb{N}.$$

That is, (f^n) is equicontinuous in $\mathbb{D}(z_0)$ considered as a family of functions defined from $\mathbb{C}$ into $\hat{\mathbb{C}}$, where $\hat{\mathbb{C}}$ is endowed with the chordal metric.

Definition 2.1.3 (Normal families with values in $\hat{\mathbb{C}}$). Let Ω be a region of $\mathbb{C}$ and let $\mathbb{F}$ be a family of analytic functions defined from Ω into $\hat{\mathbb{C}}$. We will say that $\mathbb{F}$ is normal if each

https://doi.org/10.1515/9783111689685-002

sequence $(f_n)_n$ in $\mathbb{F}$ admits a subsequence $(f_{n_k})_k$ which converges uniformly on compact subsets of Ω to some limit function.

The Arzelà–Ascoli theorem relates the notions of equicontinuity and normality. Recall that a property holds locally in Ω if it holds in a neighborhood of every point in Ω.

Theorem 2.1.4 (Arzelà–Ascoli theorem). *Let $\Omega \subset \mathbb{C}$ be a region and let $\mathbb{F}$ be a family of continuous functions $f : \Omega \to \hat{\mathbb{C}}$. The following statements are equivalent:*
(a) *Every sequence (f_n) in $\mathbb{F}$ admits a subsequence $(f_{n_k})_k$ which is locally uniformly convergent in Ω.*
(b) *The family $\mathbb{F}$ is locally d_c-equicontinuous in Ω.*

We need the following lemma.

Lemma 2.1.5. *If (f_n) is a sequence of analytic functions which converges uniformly on compact subsets of a region $\Omega \subset \mathbb{C}$ (with respect to the chordal metric), then the limit function f is analytic in Ω or $f \equiv \infty$.*

Proof. Let $f(z) = \lim_{n\to\infty} f_n(z)$ in the sense of the lemma. We will prove that, for every $z \in \Omega$, if $f(z) = \infty$, there exists a neighborhood of z where $f = \infty$ and, if $f(z) \neq \infty$, there exists a neighborhood of z where f is analytic. With this in hand, as Ω is a region, it is a standard argument to conclude the proof of the lemma. If $f(z_0) \neq \infty$, by continuity of f, there exists a disc $\mathbb{D}(z_0)$ such that f is bounded in $\overline{\mathbb{D}}(z_0)$. Since (f_n) is uniformly convergent, there exists an n_0 such that $f_n \neq \infty$ in the mentioned neighborhood for $n > n_0$. Then, the Weierstrass theorem asserts that f is analytic in a neighborhood of z_0. Now, if $f(z_0) = \infty$, by the continuity of $f : \Omega \to \hat{\mathbb{C}}$, there is a neighborhood $\overline{\mathbb{D}}(z_0)$ such that $d_c(f(z), 0) > r > 0$ for $z \in \overline{\mathbb{D}}(z_0)$. Again, by the uniform convergence, there is an n_0 such that $d_c(f_n(z), 0) > r$ for $n \geq n_0$ and $z \in \overline{\mathbb{D}}(z_0)$. Thus the f_n's ($n \geq n_0$) and f do not vanish in $\overline{D}(z_0)$, and $(1/f_n)_{n\geq n_0} \to 1/f$ uniformly on $\overline{\mathbb{D}}(z_0)$. Finally, notice that $(1/f_n)(z) \neq 0$ for z in $\overline{D}(z_0)$, while $(1/f)(z_0) = 0$. By Hurwitz's theorem, $1/f$ must be null in a neighborhood of z_0. □

In view of the above lemma, the notion of a normal family in the case of functions $f : \Omega \subset \mathbb{C} \to \mathbb{C}$ adopts the following form.

Definition 2.1.6 (Families with values in $\mathbb{C}$). Let $\Omega \subset \mathbb{C}$ be a region and let $\mathbb{F}$ be a family of analytic functions $f : \Omega \to \mathbb{C}$. We will say that $\mathbb{F}$ is normal if every sequence in $\mathbb{F}$ admits either a subsequence that converges uniformly on compact subsets of Ω, or a subsequence that diverges uniformly to ∞ on compact subsets of Ω.

Example 2.1.7. Let $f_n(z) = z+n$. Then $(f_n(z))_n$ is locally uniformly divergent in $\mathbb{C}$. In fact, given a compact set $K \subset \mathbb{C}$, we have

$$|f_n(z)| \geq \big|n - |z|\big| \geq n - M,$$

for $z \in K$ and $n \geq M$, where $M = \max\{|z| : z \in K\}$. Thus $(f_n(z))_n$ is normal in $\mathbb{C}$.

2.2 Normality conditions

We recall a classical result of Montel, the three-point condition, that we will use often.

Theorem 2.2.1 (Montel theorem). *Let $\mathbb{F}$ be a family of analytic functions with values in $\hat{\mathbb{C}}$. If there exist three different values a, b, and c in $\hat{\mathbb{C}}$ that are omitted by all $f \in \mathbb{F}$, then $\mathbb{F}$ is normal. If the functions have values in $\mathbb{C}$, $\mathbb{F}$ is normal if there are two different values a and b in $\mathbb{C}$ that are omitted.*

Now we state the properties about normality that we will need in later chapters. We only prove those results that usually are not considered in a standard course on functions of a complex variable. The proof of Marty's theorem may be consulted in the well-known book [2].

Theorem 2.2.2 (Marty's theorem). *A family $\mathbb{F}$ of analytic functions in Ω is normal if and only if for every compact $K \subset \Omega$ there exists a constant $M > 0$ such that*

$$\frac{|f'(z)|}{(1 + |f(z)|^2)} \leq M,$$

for all $z \in K$ and $f \in \mathbb{F}$.

The next theorem gives a sufficient condition of normality in terms of boundedness (Exercise 1).

Theorem 2.2.3. *If $\mathbb{F}$ is a family of analytic functions defined in Ω which is uniformly bounded on compact subsets of Ω, then $\mathbb{F}$ is normal.*

Now we will see a sharper version of Montel's theorem. Note that, in its proof, we use the classical result of Montel.

Theorem 2.2.4. *Let $\Omega \subset \mathbb{C}$ be a region and let $\mathbb{F}$ be a family of analytic functions defined on Ω. If each $f \in \mathbb{F}$ omits two different values $a_f, b_f \in \mathbb{C}$ such that*

$$|a_f|, |b_f| < M \quad \text{and} \quad |a_f - b_f| > d, \tag{2.1}$$

where $M > 0$ and $d > 0$ are constants independent of f, then $\mathbb{F}$ is normal.

Proof. We choose a sequence (f_n) in $\mathbb{F}$ and put

$$g_n(z) = \frac{f_n(z) - a_n}{b_n - a_n}.$$

Notice that each g_n does not take the values 1 and 0 in Ω. By Montel's theorem, the sequence (g_n) is normal. Then, there is a subsequence $(g_{n_k})_k$ that converges (or diverges) uniformly on compact subsets of Ω to an analytic function $g(z)$ (or ∞). In the second case, it is easy to deduce from (2.1) that $(f_{n_k})_k \to \infty$ uniformly on compact subsets of Ω. In the first case, it follows from (2.1) that $(f_{n_k})_k$ is uniformly bounded on compact subsets of Ω and, therefore, normal. Thus $(f_{n_k})_k$ admits a subsequence that is uniformly convergent on every compact set in Ω. □

We finish this chapter with Zalcman's theorem [79] that plays a fundamental role in the proof, due to Schwick (1997), of the first fundamental theorem of complex dynamics. Our proof of Zalcman's theorem is inspired by [47].

Theorem 2.2.5. *Let $\mathbb{F}$ be a family of analytic functions defined in the unit disc $\mathbb{D}$. Then $\mathbb{F}$ is nonnormal at the origin if and only if there exist sequences (f_n) in $\mathbb{F}$, (z_n) in $\mathbb{D}$ converging to zero, and (t_n) of positive numbers converging to zero such that $h_n(z_n + t_n\zeta)$ converges to a nonconstant analytic function $h(\zeta)$ uniformly on compact subsets of $\mathbb{C}$.*

Proof. (a) Suppose that $\mathbb{F}$ is not normal at $z = 0$. To apply Marty's theorem, we put

$$\kappa[f](z) = \frac{|f'(z)|}{1 + |f(z)|^2}.$$

There exist sequences (f_n) in $\mathbb{F}$ and (ζ_n) with $|\zeta_n| < (1/2n)$ satisfying

$$\kappa[f_n](\zeta_n) > 2n^2. \tag{2.2}$$

Put

$$P_n(\rho, z) = \frac{(1 - n|z|)\rho|f_n'(z)|}{1 + |f_n(z)|^2} = (1 - n|z|)\,\rho\,\kappa[f_n](z), \tag{2.3}$$

for $\rho \in [0, 1]$ and $|z| < (1/n)$. Notice that

$$0 \leq P_n(\rho, z) \leq (1 - n|z|)\rho|f_n'(z)|. \tag{2.4}$$

For every $n \in \mathbb{N}$, let $g_n(\rho) = \max\{P_n(\rho, z) : |z| \leq 1/n\}$. The functions g_n have the following properties:

(a) $g_n(1) \geq P_n(1, \zeta_n) > 1$. Indeed, by (2.2) and (2.3), we have

$$P_n(1, \zeta_n) = (1 - n|\zeta_n|)\kappa[f_n](\zeta_n) \geq n^2 > 1,$$

since $1 - n|\zeta_n| \geq 1/2$.

(b) $g_n(\rho) < 1$ for sufficiently small ρ. In fact, it follows from (2.4) that $g_n(\rho) \leq \rho M(1/n, f_n')$.

Therefore, for every $n \in \mathbb{N}$, there are ρ_n and z_n satisfying

$$1 = g_n(\rho_n) = P_n(\rho_n, z_n), \quad |z_n| < 1/n, \;\; \rho_n \in (0, 1). \tag{2.5}$$

Combining this with (2.2) and (2.3), we deduce

$$1 \geq P_n(\rho_n, \zeta_n) = (1 - n|\zeta_n|)\rho_n\kappa[f_n](\zeta_n) \geq \rho_n n^2,$$

which implies that $n\rho_n \to 0$ as $n \to \infty$.

Now we consider the function h_n defined by $h_n(\zeta) = f_n(z_n + t_n\zeta)$, where $t_n = (1 - n|z_n|)\rho_n$ for every $n \in \mathbb{N}$. Given $R > 0$, note that $|z_n + t_n\zeta| < 1/n$ for n sufficiently large and every $\zeta \in \mathbb{D}_r$. By (2.5),

$$\kappa[h_n](\zeta) = \frac{t_n|f_n'(z_n + t_n\zeta)|}{1 + |f_n(z_n + t_n\zeta)|^2} \quad \text{and} \quad \kappa[h_n](0) = 1. \tag{2.6}$$

We will prove that, for sufficiently large n, we have $1 - n|z_n + t_n\zeta| > 1$ for all $|\zeta| < R$. As $n\rho_n \to 0$ as $n \to \infty$, we choose n_0 such that $Rn\rho_n < 1/2$ for $n \geq n_0$. Thus, for those n and every $|\zeta| < R$,

$$\begin{aligned} 1 - n|z_n + t_n\zeta| &= 1 - \left|nz_n + (1 - n|z_n|)n\rho_n\zeta\right| \\ &\geq 1 - (n|z_n| + (1 - n|z_n|)n\rho_n|\zeta|) \\ &> 1 - (n|z_n| + (1 - n|z_n|)(1/2)) = 1 - \frac{1 + n|z_n|}{2} \geq 0, \end{aligned}$$

since $n|z_n| \leq 1$ for all n. Put

$$\varepsilon_n = \left|1 - \frac{1 - n|z_n + t_n\zeta|}{1 - n|z_n|}\right|.$$

It is easy to show that $0 \leq \varepsilon_n \leq Rn\rho_n$ and, therefore, $\varepsilon_n \to 0$. Obviously, we have

$$(1 - \varepsilon_n)(1 - n|z_n|) \leq 1 - n|z_n + t_n\zeta| \leq (1 - n|z_n|)(1 + \varepsilon_n).$$

Finally, we estimate $\kappa[h_n](\zeta)$. The above inequalities, together with (2.5) and (2.6), imply that

$$\begin{aligned} \kappa[h_n](\zeta) = \frac{t_n|f_n'(z_n + t_n\zeta)|}{1 + |f_n(z_n + t_n\zeta)|^2} &= \frac{1 - n|z_n|}{1 - n|z_n + t_n\zeta|} P_n(\rho_n, z_n + t_n\zeta) \\ &\leq \frac{1 - n|z_n|}{1 - n|z_n + t_n\zeta|} \leq \frac{1}{1 - \varepsilon_n}. \end{aligned}$$

Marty's criterion tells us that (h_n) is normal in $\mathbb{C}$ and, therefore, it admits a subsequence that converges uniformly on compact subsets of $\mathbb{C}$ to an analytic nonconstant function since $\kappa[h_n](0) = 1$ for all $n \in \mathbb{N}$.

(b) Suppose that $\mathbb{F}$ is normal at the origin. Given the sequences (f_n) in $\mathbb{F}$, (z_n) converging to a point $z_0 \in \mathbb{D}$ and (t_n) of positive numbers converging to 0, put $h_n(\zeta) = f_n(z_n + t_n\zeta)$. Since

$$\kappa[h_n](\zeta) = t_n\,\kappa[f_n](z_n + t_n\zeta),$$

it follows that $\kappa[h_n](\zeta)$ converges uniformly to 0 on compact subsets of $\mathbb{C}$. Then Marty's criterion asserts that (h_n) is normal in $\mathbb{C}$. Finally, assume that $h = \lim_{n\to\infty} h_n$ uniformly on compact subsets of $\mathbb{C}$. Note that $\kappa[h](\zeta) = \lim_{n\to\infty} \kappa[h_n](\zeta) = 0$ and this yields $h' \equiv 0$. □

2.3 Exercises

1. Prove the easy part of Marty's theorem: If $\mathbb{F}$ is a normal family of analytic functions in a region $\Omega \subset \mathbb{C}$, then $|f'(z)|/(1+|f(z)|^2)$ is uniformly bounded on every compact $K \subset \Omega$.
2. Prove Theorem 2.2.3: If $\mathbb{F}$ is a family of analytic functions defined in a region $\Omega \subset \mathbb{C}$ which is uniformly bounded in compact subsets of Ω, then $\mathbb{F}$ is normal.
3. Let (f_n) be a sequence of analytic functions defined in a region $\Omega \subset \mathbb{C}$. If (f_n) is normal and pointwise convergent on Ω, then the convergence is uniform on every compact.
4. If $E_\lambda(z) = \lambda e^z$, prove that $\lambda(1+z/N)^N \to E_\lambda(z)$ as $n \to \infty$ uniformly on every compact subset of $\mathbb{C}$.
5. Let $\Omega \subset \mathbb{C}$ be a bounded region, $z_0 \in \Omega$, and $\mathbb{F}$ the family of analytic functions given by

$$\mathbb{F} = \{f \in H(\Omega) : f(\Omega) \subset \mathbb{D}, f(z_0) = f'(z_0) = 0, f''(z_0) > 0\}.$$

   (i) Show that $\mathbb{F}$ contains a polynomial of degree 2.
   (ii) If $\alpha = \sup\{f''(z_0) : f \in \mathbb{F}\}$, prove that α is finite and there is an $f \in \mathbb{F}$ such that $f''(z_0) = \alpha$.
   (iii) Now suppose that $\Omega = \mathbb{D}$ and $z_0 = 0$. Prove that $|f(z)| \le |z|^2$ for all $f \in \mathbb{F}$ and $z \in \mathbb{D}$, and deduce that $\alpha = 2$.
6. Let (a_n), (b_n) be two sequences in $(0, +\infty)$ such that (a_n) is increasing and $a_n \to 1$. If $\mathbb{F}$ is the family of analytic functions given by

$$\mathbb{F} = \left\{f \in H(\mathbb{D}) : |f(0)| \le 1, \sup_{|z|=a_n} |f'(z)| \le b_n \ (\forall n \in \mathbb{N})\right\},$$

   prove that $\mathbb{F}$ is normal.

Bibliography remarks

Section 2.1 is based on [2, Section V.5].

In Section 2.2, we follow [47, Section 1.2].

3 Value distribution theory

The theory of the distribution of values of meromorphic functions, developed by R. Nevanlinna and L. Ahlfors, may be consider as one of the most relevant achievements of mathematics in the last century. We include the proofs, among other results, of the two fundamental theorems of Nevanlinna. However, some secondary results that we use in the iteration theory of entire functions only will be proved for the case of entire functions. For a more comprehensive study, we refer the reader to the monograph of Hayman [42].

3.1 The Poisson–Jensen formula

If f is analytic in Ω, recall that the real part of f is harmonic in Ω. Given an analytic function f, we will see that $\log|f(z)|$ is harmonic except at the zeros of f. Indeed, if $f(z_0) \neq 0$, take a disc $\mathbb{D}(z_0)$ with center z_0 such that $f \neq 0$ in $\mathbb{D}(z_0)$. As $\mathbb{D}(z_0)$ is simply connected, there is an analytic branch of $\log f(z)$ defined in $\mathbb{D}(z_0)$ and $\log|f(z)|$ is precisely its real part. Then, if f is analytic and does not have zeros in $|z| \leq R$, the Poisson formula yields

$$\log|f(z)| = \frac{1}{2\pi} \int_0^{2\pi} \frac{R^2 - r^2}{R^2 - 2rR\cos(\phi - \theta) + r^2} \log|f(Re^{i\theta})|\, d\theta, \tag{3.1}$$

for all $z = re^{\phi i}$, with $0 \leq r < R$. We will need the transformations $w = (R^2 - \overline{a}z)/(R(z - a))$, where $|a| < R$. Elementary calculus shows that $|w| = 1$ is equivalent to $|z| = R$.

Theorem 3.1.1 (Poisson–Jensen's formula). *Let f be a nonconstant meromorphic function in $|z| \leq R$. We denote by $\{a_j\}_{j\in J}$ and $\{b_k\}_{k\in K}$ the zeros and poles, respectively, of f in $|z| < R$ (zeros or poles of order p are repeated p times and ordered such that their modulus are nondecreasing). We suppose that $a_1 \neq 0$ and $b_1 \neq 0$ and there are no zeros or poles in $|z| = R$. Then, if $z = re^{\phi i}$ $(0 \leq r < R)$, we have*

$$\begin{aligned} \log|f(z)| = {} & \frac{1}{2\pi} \int_0^{2\pi} \frac{R^2 - r^2}{R^2 - 2rR\cos(\phi - \theta) + r^2} \log|f(Re^{i\theta})|\, d\theta \\ & + \sum_{j\in J} \log\left|\frac{R(z - a_j)}{R^2 - \overline{a}_j z}\right| - \sum_{k\in K} \log\left|\frac{R(z - b_k)}{R^2 - \overline{b}_k z}\right|. \end{aligned} \tag{3.2}$$

Proof. Consider the function

$$F(z) = f(z)\left[\prod_{k\in K} \frac{R(z - b_k)}{R^2 - \overline{b}_k z}\right]\left[\prod_{j\in J} \frac{R^2 - \overline{a}_j z}{R(z - a_j)}\right].$$

https://doi.org/10.1515/9783111689685-003

Then F is analytic and has no zeros in $|z| \leq R$. As $|F(z)| = |f(z)|$ for all $|z| = R$, applying (3.1) to F, concludes the proof. □

The Poisson–Jensen's formula has a simpler form in the case $z = 0$, namely

$$\log|f(0)| = \frac{1}{2\pi}\int_0^{2\pi} \log|f(Re^{i\theta})|\,d\theta + \sum_{|a_j|<R} \log\frac{|a_j|}{R} - \sum_{|b_k|<R} \log\frac{|b_k|}{R}. \tag{3.3}$$

We will often use this formula and, as usual, we will refer to it as Jensen's formula.

Remark 3.1.2. If the origin is a pole of f, Jensen's formula has the form

$$\log|c| = \frac{1}{2\pi}\int_0^{2\pi} \log|f(Re^{i\theta})|\,d\theta + m\log R + \sum_{|a_j|<R} \log\frac{|a_j|}{R} - \sum_{|b_k|<R} \log\frac{|b_k|}{R}. \tag{3.4}$$

In fact, if we apply the above theorem to the function $F(z) = f(z)z^m$, where m is the order of the pole and $c = \lim_{z\to 0} f(z)z^m$, we obtain

$$\log|F(0)| = \frac{1}{2\pi}\int_0^{2\pi} \log|R^m f(Re^{i\theta})|\,d\theta + \sum_{|a_j|<R} \log\frac{|a_j|}{R} - \sum_{|b_k|<R} \log\frac{|b_k|}{R}.$$

As $F(0) = \lim_{z\to 0} f(z)z^m$, we have proved (3.4).

Now we will express Jensen's formula in terms of the function that counts the number of poles of f. Concretely, we denote by $n(t,f)$ the number of poles of f in $|z| \leq t$, counting multiplicities. Then the last term on the right-hand side of (3.4) may be modified as follows:

$$-\sum_{|b_k|<R} \log(|b_k|/R) = \int_0^R -\log(t/R)\,d\alpha(t),$$

where $\alpha(t) = n(t,f) - n(0,f)$. Note that $\alpha = 0$ in $[0,r]$, with r being the modulus of the first nonzero pole of f. Thus, integrating by parts, we obtain

$$-\sum_{|b_k|<R} \log(|b_k|/R) = -[\alpha(t)\log(t/R)]_0^R + \int_0^R \alpha(t)d\log(t/R) = \int_0^R \frac{\alpha(t)}{t}\,dt.$$

Therefore, in Jensen's formula (3.4), the part related to the poles of f in $|z| < R$ has the final form

$$n(0,f)\log R - \sum_{|b_k|<R} \log(|b_k|/R) = \int_0^R \frac{n(t,f) - n(0,f)}{t}\,dt + n(0,f)\log R.$$

This expression allows us to interpret the left-hand side as a mean of values of $n(t,f)$. This is the reason why the function $N(r,f)$ given by

$$N(r,f) = \int_0^r \frac{n(t,f) - n(0,f)}{t}\, dt + n(0,f)\log r$$

is called the *counting function* (for the poles) of f.

If f has a zero of order m at the origin, we can obtain Jensen's formula in a similar way:

$$\log|c| = \frac{1}{2\pi}\int_0^{2\pi} \log|f(Re^{i\theta})|\, d\theta - m\log R + \sum_{|a_j|<R} \log\frac{|a_j|}{R} - \sum_{|b_k|<R} \log\frac{|b_k|}{R}, \tag{3.5}$$

where $c = \lim_{z\to 0} f(z)z^{-m}$. Both formulas, (3.4) and (3.5), admit the following representation:

$$\log|c| = \frac{1}{2\pi}\int_0^{2\pi} \log|f(Re^{i\theta})|\, d\theta + \epsilon m\log R + \sum_{|a_j|<R} \log\frac{|a_j|}{R} - \sum_{|b_k|<R} \log\frac{|b_k|}{R}, \tag{3.6}$$

where $\epsilon = 1$ (resp. $\epsilon = -1$), if the origin is a pole (resp. zero) of f of order m, and $c = \lim_{z\to 0} f(z)z^m$ (resp. $c = \lim_{z\to 0} f(z)z^{-m}$), if the origin is a pole (resp. zero). In terms of the counting function of poles, the equality (3.6) becomes

$$\log|c| = \frac{1}{2\pi}\int_0^{2\pi} \log|f(Re^{i\theta})|\, d\theta + N(R,f) - N(R,1/f), \tag{3.7}$$

in the case that f has a pole of order m at the origin.

3.2 The Nevanlinna characteristic function

Recall that $\log^+ x = \max\{\log x, 0\}$. We will need some basic properties of $\log^+ x$:
1. $\log x = \log^+ x - \log^+(1/x)$, for all $x > 0$.
2. If $0 < x \le y$, then $\log^+ x \le \log^+ y$.
3. $\log^+ \prod_{i=1}^n |a_i| \le \sum_{i=1}^n \log^+ |a_i|$.

If $\prod_{i=1}^n |a_i| < 1$, it is obvious. Otherwise, we have

$$\log^+ \prod_{i=1}^n |a_i| = \log \prod_{i=1}^n |a_i| = \sum_{i=1}^n \log|a_i| \le \sum_{i=1}^n \log^+ |a_i|.$$

The inequality in the last step occurs because some a_i may have modulus less than 1 and, in this case, $\log|a_i|$ is negative, while $\log^+|a_i|$ is zero.

4. $\log^+|\sum_{i=1}^n a_i| \le \log^+(n \max|a_i|) \le \sum_{i=1}^n \log^+|a_i| + \log n$.

In view of the inequalities $|\sum_{i=1}^n a_i| \le \sum_{i=1}^n |a_i| \le n \max|a_i|$, it follows that

$$\log^+\left|\sum_{i=1}^n a_i\right| \le \log^+(n \max|a_i|) \le \sum_{i=1}^n \log^+|a_i| + \log n.$$

Definition 3.2.1. If f is a nonconstant meromorphic function in $\mathbb{C}$, the function defined by

$$m(r,f) = \frac{1}{2\pi}\int_0^{2\pi} \log^+|f(re^{i\theta})|\, d\theta$$

is called the proximity function of f.

The function $T(r,f) = m(r,f) + N(r,f)$ is called the characteristic function of f (in the sense of Nevanlinna).

By property (1) of $\log^+$, it is easy to deduce the equality

$$m(r,f) - m(r,1/f) = (1/2\pi)\int_0^{2\pi} \log|f(re^{i\theta})|\, d\theta.$$

On the other hand, if the origin is a pole of f of order m, it follows from (3.7) that

$$T(r,f) - T(r,1/f) = \log|c|, \tag{3.8}$$

where $c = \lim_{z\to 0} f(z)z^m$.

Examples 3.2.2. (1) If f is a rational function of degree d, then $T(r,f) = d\log r + O(1)$.

Let $f = P/Q$ with P and Q polynomials of degree p and q, respectively (we suppose that P and Q have no common zeros). Recall that the degree of f is $d = \max\{p,q\}$. First, note that f has a finite number of poles (the zeros of Q) and, therefore, $n(r,f) = q$ for $r > r_0$. Then it is easy to deduce that $N(r,f) = q\log r + O(1)$.

Now we consider the case $p < q$. As $\lim_{z\to\infty} f(z) = 0$, there exists an $r_0 > 0$ such that $|f(z)| < 1$ for $|z| > r_0$. Then $m(r,f) = 0$ for $r > r_0$ and, in consequence, we have

$$T(r,f) = N(r,f) = q\log r + O(1).$$

In the case $p > q$, it follows from (3.8) that

$$T(r,f) = T(r,1/f) + O(1) = p\log r + O(1).$$

Finally, consider the case $p = q$. As $\lim_{z\to\infty} |f(z)| = M < +\infty$, for each $\varepsilon > 0$, there exists an $r_0 > 0$ such that

$$M - \varepsilon < |f(z)| < M + \varepsilon \quad \text{for all } |z| > r_0.$$

Taking the logarithm and integrating on $[0, 2\pi]$, we obtain

$$\log^+(M - \varepsilon) \le \log^+|f(z)| \le \log^+(M + \varepsilon) \quad \text{for all } |z| > r_0.$$

Then $\log^+(M-\varepsilon) \le m(r,f) \le \log^+(M+\varepsilon)$ for $r > r_0$ and $\varepsilon > 0$. Therefore, $m(r,f) = \log^+ M$ for $r > r_0$, and we conclude that $T(r,f) = q \log r + O(1)$.

(2) If $f(z) = a$ for every $z \in \mathbb{C}$, then $T(r,f) = \log^+ |a|$.

The following properties of the functions $N(r,f)$, $m(r,f)$, and $T(r,f)$ will be used repeatedly in the proof of the fundamental theorems of Nevanlinna. We suppose that the functions f_i are meromorphic.

Properties (3) and (4) of $\log^+$ and the definition of $m(r,f)$ yield in an obvious way:
(i) $m(r, \sum_{i=1}^p f_i) \le \sum_{i=1}^p m(r,f_i) + \log p$.
(ii) $m(r, \prod_{i=1}^p f_i) \le \sum_{i=1}^p m(r,f_i)$.
(iii) $N(r, \sum_{i=1}^p f_i) \le \sum_{i=1}^p N(r,f_i)$.
(iv) $N(r, \prod_{i=1}^p f_i) \le \sum_{i=1}^p N(r,f_i)$.

If f is the sum or the product of the functions f_i, the order of a pole z_0 of f is at most the sum of the orders of z_0 as a pole of f_i. Then we have

$$n(r,f) \le \sum_i n(r,f_i) \quad \text{and} \quad n(r,f) - n(0,f) \le \sum_i (n(r,f_i) - n(0,f_i)).$$

This, together with the definition of $N(r,f)$, easily gives (iii) and (iv).

Finally, using the above properties and the definition of $T(r,f)$, we deduce:
(v) $T(r, \sum_{i=1}^p f_i) \le \sum_{i=1}^p T(r,f_i) + \log p$.
(vi) $T(r, \prod_{i=1}^p f_i) \le \sum_{i=1}^p T(r,f_i)$.
(vii) $|T(r,f) - T(r,f-a)| \le \log^+ |a| + \log 2$.

Let us see the proof of (vii). Applying (v) to $f_1(z) = f(z)$ and $f_2(z) = -a$, we have

$$T(r,f-a) \le T(r,f) + T(r,-a) + \log 2 = T(r,f) + \log^+ |a| + \log 2.$$

Again applying (v) with $f_1 = f - a$ and $f_2 = a$, we obtain

$$T(r,f) \le T(r,f-a) + T(r,a) + \log 2 = T(r,f-a) + \log^+ |a| + \log 2,$$

which yields (vii).

Theorem 3.2.3 (The first fundamental theorem of Nevanlinna). *For every nonconstant meromorphic function f and every $a \in \mathbb{C}$, the following statement holds:*

$$T(R, 1/(f-a)) = T(R,f) - \log|f(0)-a| + \epsilon(a,R),$$

where

$$|\epsilon(a,R)| \le \log^+|a| + \log 2.$$

Proof. To simplify the proof, we suppose that $a \neq f(0) \neq \infty$. By (3.8), we have

$$T\left(R, \frac{1}{f-a}\right) = T(R, f-a) + \log|f(0)-a|.$$

Adding and subtracting $T(R,f)$, we obtain

$$T\left(R, \frac{1}{f-a}\right) = T(R, f-a) - T(R,f) + T(R,f) + \log|f(0)-a|.$$

Putting $\epsilon(R,a) = T(R, f-a) - T(R,f)$, the proof concludes applying (vii). □

Sometimes, to simplify, we will write $m(r,a)$, $N(r,a)$, $n(r,a)$, and $T(r)$ instead of $m(r,1/(f-a))$, $N(r,1/(f-a))$, $n(r,1/(f-a))$, and $T(r,f)$ if a is finite; and $m(r,\infty)$, $N(r,\infty)$, and $n(r,\infty)$ instead of $m(r,f)$, $N(r,f)$, and $n(r,f)$ (if there is no doubt as to which function f one refers).

With this notation, the first fundamental theorem may be written in the form

$$m(r,a) + N(r,a) = T(r,f) + O(1)$$

for every a, finite or infinite. The term $N(r,a)$ refers to the number of roots of the equation $f(z) = a$ in $|z| < r$, while the term $m(r,a)$ is, in a certain sense, the average value of $f(z) - a$ in the circle $|z| = r$ (more precisely, in the part of the circle where $|f(z) - a| \ge 1$). The first fundamental theorem asserts that, for every a, the sum of these two terms is the same apart from a bounded term. However, we will see later that, in general, $N(r,a)$ dominates.

Example 3.2.4. If f is a polynomial of degree p, the equation $f(z) = a$ has p roots. Then $n(r,a) = p$ for $r > r_0$ and, therefore, $N(r,a) = p\log r + O(1)$ for $r > r_0$. On the other hand, $m(r,a) = 0$ for r sufficiently large. In fact, as $\lim_{z\to\infty} f(z) = \infty$, we may choose an r_0 such that $|f(z)| > 1 + |a|$ for $r > r_0$. Then we have

$$\frac{1}{|f(z)-a|} \le \frac{1}{|f(z)| - |a|} < 1,$$

which implies that $m(r,a) = 0$ for $r > r_0$. Thus the first fundamental theorem tells us that

$$T(r) = N(r,a) + \log|f(0)-a| + \epsilon(a,r) = p\log r + O(1),$$

for $r > r_0$.

The next identity was proved by Nevanlinna (1929). Here we follow the proof of Cartan.

Theorem 3.2.5 (Cartan's formula). *If f is meromorphic in $|z| < R$, then*

$$T(r,f) = \frac{1}{2\pi}\int_0^{2\pi} N\left(r, \frac{1}{f - \exp(\theta i)}\right) d\theta + \log^+|f(0)| \quad \textit{for all } 0 < r < R.$$

Proof. Applying Jensen's formula to the function $g(z) = a - z$ with $R = 1$, we obtain

$$\log|a| = \frac{1}{2\pi}\int_0^{2\pi} |a - e^{\theta i}|\, d\theta + A,$$

where $A = 0$, if $|a| \geq 1$, and $A = \log|a|$ in the case $|a| < 1$. Then in all cases,

$$\log^+|a| = \frac{1}{2\pi}\int_0^{2\pi} |a - e^{\theta i}|\, d\theta. \tag{3.9}$$

We again apply Jensen's formula, but this time to the function $f(z) - e^{\theta i}$ and get

$$\log|f(0) - e^{\theta i}| = \frac{1}{2\pi}\int_0^{2\pi} \log^+|f(re^{\phi i}) - e^{\theta i}|\, d\phi + N(r,f) - N(r, 1/(f - e^{\theta i})).$$

Integrating with respect to θ and changing the order of integration in the double integral, we obtain

$$\begin{aligned}\frac{1}{2\pi}\int_0^{2\pi} \log|f(0) - e^{\theta i}|\, d\theta &= \frac{1}{2\pi}\int_0^{2\pi}\left[\frac{1}{2\pi}\int_0^{2\pi} \log|f(re^{\phi i}) - e^{\theta i}|\, d\theta\right] d\phi \\ &\quad + N(r,f) - \frac{1}{2\pi}\int_0^{2\pi} N(r, 1/(f - e^{\theta i}))\, d\theta.\end{aligned}$$

Finally, using (3.9), we deduce

$$\begin{aligned}\log^+|f(0)| &= \frac{1}{2\pi}\int_0^{2\pi} \log^+|f(re^{\phi i})|\, d\phi + N(r,f) - \frac{1}{2\pi}\int_0^{2\pi} N(r, 1/(f - e^{\theta i}))\, d\theta \\ &= T(r,f) - \frac{1}{2\pi}\int_0^{2\pi} N(r, 1/(f - e^{\theta i}))\, d\theta.\end{aligned}$$

□

Theorem 3.2.6. *Let f be analytic in $|z| \leq R$ and $M(r,f) = \max\{|f(z)| : |z| = r\}$. The following inequalities hold:*

$$T(r,f) \le \log^+ M(r,f) \le \left(\frac{R+r}{R-r}\right) T(R,f),$$

for all $0 < r < R$.

Proof. As f is analytic, it follows that $N(r,f) = 0$ for all $0 < r \le R$. Thus

$$T(r,f) = m(r,f) = \frac{1}{2\pi}\int_0^{2\pi} \log^+|f(re^{\theta i})|\, d\theta \le \log^+ M(r,f).$$

If $M(r,f) \le 1$, the second inequality is evident. Then we assume that $M(r,f) \ge 1$ and choose $z_0 = re^{\phi i}$ such that $M(r,f) = |f(z_0)|$. First, we consider the case that f does not have any zero in $|z| < R$. By Poisson formula,

$$\begin{aligned}\log^+ M(r,f) = \log|f(re^{\phi i})| &= \frac{1}{2\pi}\int_0^{2\pi} \frac{(R^2 - r^2)\log^+|f(Re^{\theta i})|}{R^2 - 2rR\cos(\phi - \theta) + r^2}\, d\theta \\ &\le \left(\frac{R+r}{R-r}\right)\frac{1}{2\pi}\int_0^{2\pi} \log^+|f(Re^{\theta i})|\, d\theta = \left(\frac{R+r}{R-r}\right) T(R,f).\end{aligned}$$

Now we suppose that $\{a_j : j \in J\}$ are the zeros of f in $|z| < R$ (zeros of order p are repeated p times). Consider the function

$$F(z) = f(z)\left[\prod_{j\in J} \frac{R^2 - \bar{a}_j z}{R(z - a_j)}\right].$$

Both functions F and f are analytic, therefore we have $N(r,f) = N(r,F) = 0$. On the other hand, $m(R,f) = m(R,F)$ and $M(R,F) = M(R,f)$ since $|F(z)| = |f(z)|$ in $|z| = R$. As F does not have any zero in $|z| < R$, we have

$$\log^+ M(r,f) \le \log^+ M(r,F) \le \left(\frac{R+r}{R-r}\right) T(R,F),$$

and the proof concludes. □

Theorem 3.2.7. *A function f is a transcendental entire function if and only if*

$$\lim_{r\to+\infty} \frac{T(r,f)}{\log r} = \infty.$$

Proof. Applying the above result with $r = R/2$, we obtain

$$\log^+ M(R/2,f) \le 3T(R,f).$$

Then, if f is transcendental, we know that $\lim_{r\to+\infty} M(r,f)/\log r = +\infty$ and the conclusion follows. Finally, if f is a polynomial of degree p, Example 3.2.4 tells us that $\lim_{r\to+\infty} T(r,f)/\log r = p$. □

3.3 The second fundamental theorem of Nevanlinna

Theorem 3.3.1 (The second fundamental theorem of Nevanlinna). *Suppose that f is a nonconstant meromorphic function. If $q \geq 2$ and $a_1,\ldots,a_q$ are different complex numbers, the following statements hold:*

(i)

$$m(r,\infty) + \sum_{j=1}^{q} m(r,a_j) \leq 2T(r,f) - N_1(r) + S(r,f), \tag{3.10}$$

where $N_1(r) = N(r,1/f') + 2N(r,f) - N(r,f')$, and

$$S(r,f) = m(r,f'/f) + m\left(r, \sum_{\mu=1}^{q} f'(f-a_\mu)\right) + \log\frac{1}{|f'(0)|} + q\log^+\frac{3q}{\delta} + \log 2.$$

(ii) $S(r,f) = o(T(r,f))$ *(if f is of infinite order, for r outside a set E of finite length; E depends on f but not on the a_j's or q).*

Proof. The proof is quite laborious and we only prove (i). The interested reader may consult [42].

Consider the function $F(z) = \sum_{j=1}^{q}\frac{1}{f(z)-a_j}$ and put $\delta = \min\{|a_i - a_j| : i \neq j\}$. Suppose that there is some ν such that $|f(z) - a_\nu| < \delta/3q$ (note that there can only be one ν with this property). Then for $\mu \neq \nu$, we have

$$|f(z) - a_\mu| \geq |a_\mu - a_\nu| - |f(z) - a_\nu| \geq \delta - \delta/3q \geq (2/3)\delta$$

and, therefore,

$$\frac{1}{|f(z)-a_\mu|} \leq \frac{3}{2\delta} \leq \frac{1}{2q}\left(\frac{1}{|f(z)-a_\nu|}\right). \tag{3.11}$$

Using (3.11), we obtain

$$|F(z)| \geq \frac{1}{|f(z)-a_\nu|} - \sum_{\mu\neq\nu}\frac{1}{|f(z)-a_\mu|} \geq \frac{1}{2|f(z)-a_\nu|}.$$

Hence

$$\log^+|F(z)| \geq \log^+\frac{1}{|f(z)-a_\nu|} - \log 2. \tag{3.12}$$

By (3.11), $\sum_{\mu\neq\nu}\log^+\frac{1}{|f(z)-a_\mu|}-(q-1)\log^+(3/2\delta)\leq 0$. Then, adding this expression to the right-hand side of (3.12), we deduce that

$$\begin{aligned}\log^+|F(z)| &\geq \sum_{\mu=1}^{q}\log^+\frac{1}{|f(z)-a_\mu|}-(q-1)\log^+(3/2\delta)-\log 2\\ &\geq \sum_{\mu=1}^{q}\log^+\frac{1}{|f(z)-a_\mu|}-q\log^+(3q/\delta)-\log 2.\end{aligned}$$

This relation is valid for z such that $|f(z)-a_\nu|<\delta/(3q)$ for some $\nu\leq q$. If $|f(z)-a_\nu|\geq\delta/(3q)$ for all $\mu\leq q$, then it is evident that

$$\log^+|F(z)|\geq\sum_{\mu=1}^{q}\log^+\frac{1}{|f(z)-a_\mu|}-q\log^+(3q/\delta)-\log 2. \tag{3.13}$$

In fact, note that for every $\nu\leq q$,

$$\log^+\frac{1}{|f(z)-a_\mu|}\leq\log^+\frac{3q}{\delta},$$

from which we deduce

$$\sum_{\mu=1}^{q}\log^+\frac{1}{|f(z)-a_\mu|}-q\log^+(3q/\delta)\leq 0.$$

Now (3.13) follows obviously.

Integrating (3.13) on $[0,2\pi]$, we obtain

$$m(r,F)\geq\sum_{\mu=1}^{q}m(r,a_\mu)-q\log^+\frac{3q}{\delta}-\log 2$$

or, equivalently,

$$\sum_{\mu=1}^{q}m(r,a_\mu)\leq m(r,F)+q\log^+\frac{3q}{\delta}+\log 2. \tag{3.14}$$

On the other hand, for $m(r,F)$ we have

$$m(r,F)=m\left(r,\frac{1}{f}\frac{f}{f'}f'F\right)\leq m\left(r,\frac{1}{f}\right)+m\left(r,\frac{f}{f'}\right)+m(r,f'F).$$

By the first fundamental theorem,

$$m(r,1/f)=T(r,f)-N(r,1/f)+\log\frac{1}{|f(0)|}$$

and

$$m\left(r, \frac{f}{f'}\right) = T\left(r, \frac{f'}{f}\right) - N\left(r, \frac{f}{f'}\right) + \log \frac{|f(0)|}{|f'(0)|}.$$

Finally, we deduce that

$$m(r,F) \leq T(r,f) - N(r,1/f) + m\left(r, \frac{f'}{f}\right) + N\left(r, \frac{f'}{f}\right) - N\left(r, \frac{f}{f'}\right) + m(r,f'F) + \log \frac{1}{|f'(0)|}.$$

Combining with (3.14), we conclude

$$\begin{aligned}\sum_{\mu=1}^{q} m(r,a_\mu) \leq T(r,f) - N(r,1/f) + m\left(r, \frac{f'}{f}\right) + N\left(r, \frac{f'}{f}\right) - N\left(r, \frac{f}{f'}\right)\\ + m(r,f'F) + \log \frac{1}{|f'(0)|} + q\log^+ \frac{3q}{\delta} + \log 2.\end{aligned}$$

Adding $m(r,f)$ to both sides of the above inequality, we obtain

$$\begin{aligned}m(r,\infty) + \sum_{\mu=1}^{q} m(r,a_\mu) \leq T(r,f) - N(r,1/f) + N\left(r, \frac{f'}{f}\right) - N\left(r, \frac{f}{f'}\right)\\ + m\left(r, \frac{f'}{f}\right) + m(r,f'F) + T(r,f) - N(r,f)\\ + \log \frac{1}{|f'(0)|} + q\log^+ \frac{3q}{\delta} + \log 2.\end{aligned}$$

So, if we show that

$$N(r,f'/f) - N(r,f/f') = N(r,1/f) - N(r,f) - N(r,1/f') + N(r,f'),$$

the above inequality adopts the required form. In fact, by Jensen's formula (3.7), we have

$$\begin{aligned}N(r,f'/f) - N(r,f/f') &= \frac{1}{2\pi}\int_0^{2\pi} \log\left|\frac{f(re^{i\theta})}{f'(re^{i\theta})}\right| d\theta - \log\left|\frac{f(0)}{f'(0)}\right|\\ &= \frac{1}{2\pi}\int_0^{2\pi} \log|f(re^{i\theta})|\, d\theta - \log|f(0)|\\ &\quad - \frac{1}{2\pi}\int_0^{2\pi} \log|f'(re^{i\theta})|\, d\theta + \log|f'(0)|\\ &= N\left(r, \frac{1}{f}\right) - N(r,f) - N\left(r, \frac{1}{f'}\right) + N(r,f').\end{aligned}$$

□

Now, to finish the proof of the fundamental inequality of Nevanlinna, we will express the second fundamental theorem in a more convenient form. We denote by $\overline{n}(r,f)$ the number of poles of f in $|z| \le r$, disregarding multiplicity, and define $\overline{N}(r,f)$ by

$$\overline{N}(r,f) = \int_0^r \frac{\overline{n}(t,f) - \overline{n}(0,f)}{t}\,dt + \overline{n}(0,f)\log r.$$

Theorem 3.3.2. *Under the conditions of the above theorem, the following inequality holds:*

$$(q-1)T(r,f) \le \sum_{j=1}^{q} \overline{N}(r,1/(f-a_j)) + \overline{N}(r,f) + o(T(r,f)). \tag{3.15}$$

Proof. Adding $N(r,f) + \sum_{j=1}^{q} N(r,1/(f-a_j))$ to both sides of (3.10) and having in mind the first fundamental theorem, we obtain

$$(q+1)T(r,f) + O(1) \le 2T(r,f) + \sum_{j=1}^{q} N(r,1/(f-a_j)) + N(r,f) - N_1(r) + S(r,f).$$

Cancelling terms, we deduce

$$(q-1)T(r,f) + O(1) \le \sum_{j=1}^{q} N(r,1/(f-a_j)) + N(r,f) - N_1(r) + S(r,f). \tag{3.16}$$

By the definition of $N_1(r)$,

$$N(r,f) - N_1(r) = -N(r,1/f') - N(r,f) + N(r,f').$$

Now note that if z_0 is a pole of order p for f, then z_0 is a pole of f' of order $p+1$. Thus $n(r,f') - n(r,f) = \overline{n}(r,f)$ and, in consequence, $N(r,f') - N(r,f) = \overline{N}(r,f)$. Hence (3.16) may be written in the form

$$(q-1)T(r,f) + O(1) \le \sum_{j=1}^{q} N(r,1/(f-a_j)) + \overline{N}(r,f) - N(r,1/f') + S(r,f). \tag{3.17}$$

On the other hand, denote by $n_0(t,1/f')$ the number of zeros of f' in $|z| \le t$ which occur at points other than the roots of the equation $f(z) = a_j$ ($j = 1,\dots,q$), counting multiplicities. The difference $\sum_{j=1}^{q} N(r,1/(f-a_j)) - N(r,1/f')$ is the same as $\sum_{j=1}^{q} \overline{N}(r,1/(f-a_j)) - N_0(r,1/f')$ (note that the roots or order $p > 1$ for $f(z) = a_j$ are of order $p-1$ for f'). Finally, since $N_0(r,1/f') \ge 0$ (for $r > r_0$), we can write (3.17) in the form

$$(q-1)T(r,f) + O(1) \le \sum_{j=1}^{q} \overline{N}(r,1/(f-a_j)) + \overline{N}(r,f) + S(r,f), \tag{3.18}$$

which is equivalent to (3.15). □

Remark 3.3.3. The first fundamental theorem establishes that $m(r,a) + N(r,a) = T(r,f) + O(1)$, for every complex a. Then the sum $m(r,a) + N(r,a)$ is largely independent of a. The second fundamental theorem asserts that $N(r,a)$ is the term which is dominant in that sum and, further, that in $N(r,a)$ the sum does not decrease much if multiple roots are simply counted. Then most of these roots are simple.

As a direct consequence of (3.15), we mention the well-known Picard theorem: "Every transcendental meromorphic (entire, respectively) function in the plane takes every value infinitely many times except for at most two (one, respectively)". In fact, if f is entire, taking $q = 2$, (3.15) reduces to

$$T(r,f) \le \overline{N}(r,1/(f-a_1)) + \overline{N}(r,1/(f-a_2)) + o(T(r,f)).$$

By contradiction, suppose that there are two different values a_1 and a_2 in $\mathbb{C}$ such that the equation $f(z) - a_j = 0$ has a finite number p_j of solutions for $j = 1,2$. Notice that $\overline{N}(r,1/(f-a_j)) = p_j \log r + O(1)$. Then the above inequality may be expressed in the form

$$T(r,f) \le p_1 \log r + p_2 \log r + o(T(r,f))$$

and this is in contradiction with Theorem 3.2.7.

To establish the fundamental inequality of Nevanlinna, we introduce the following three functions. Given a nonconstant meromorphic function in the plane, for each $a \in \hat{\mathbb{C}}$, we put

$$\begin{aligned}
\delta(a) &= \delta(a,f) = 1 - \varlimsup_{r\to+\infty} \frac{N(r,a)}{T(r,f)},\\
\Theta(a) &= \Theta(a,f) = 1 - \varlimsup_{r\to+\infty} \frac{\overline{N}(r,a)}{T(r,f)},\\
\theta(a) &= \theta(a,f) = \varliminf_{r\to+\infty} \frac{N(r,a) - \overline{N}(r,a)}{T(r,f)}.
\end{aligned}$$

By the definitions of $\delta(a)$ and $\theta(a)$, given $\epsilon > 0$, there exists an $R > 0$ such that

$$N(r,a) - \overline{N}(r,a) > [\theta(a) - \epsilon]T(r,f) \quad \text{and} \quad N(r,a) < [1 - \delta(a) + \epsilon]T(r,f)$$

for $r > R$, and then

$$\overline{N}(r,a) < [1 - \delta(a) - \theta(a) + 2\epsilon]T(r,f),$$

which yields

$$\Theta(a) \ge \delta(a) + \theta(a). \tag{3.19}$$

The term $\delta(a)$ is called the deficiency of the value a and $\theta(a)$ is called the index of multiplicity. Note that $\delta(a)$ is positive only if the equation $f(z) - a = 0$ has relatively few roots and $\theta(a)$ is positive if there are relatively many multiple roots.

The following fundamental result is called the Nevanlinna theorem for deficient values.

Theorem 3.3.4. *Let f be nonconstant and meromorphic in $\mathbb{C}$. The set of all values a for which $\Theta(a) > 0$ is countable and*

$$\sum_a (\delta(a) + \theta(a)) \le \sum_a \Theta(a) \le 2 \quad \textit{(Nevanlinna's inequality)}.$$

Proof. We will apply the second fundamental theorem in the form (3.15). Given q different complex numbers $a_1, \dots, a_q$, we choose $r_n \to +\infty$ such that

$$\lim_{n\to\infty} \frac{o(T(r_n,f))}{T(r_n,f)} = 0$$

and put $\alpha_n = o(T(r_n,f))/T(r_n,f)$. Then (3.15) becomes

$$\sum_{j=1}^{q} \overline{N}(r_n, 1/(f - a_j)) + \overline{N}(r_n,f) \ge (q - 1 - \alpha_n)T(r_n,f).$$

Dividing by $T(r_n,f)$, we deduce

$$\begin{aligned}
\sum_{j=1}^{q} \varliminf_{r\to+\infty} \frac{\overline{N}(r, 1/(f - a_j))}{T(r,f)} &+ \varliminf_{r\to+\infty} \frac{\overline{N}(r,f)}{T(r,f)} \\
&\ge \varliminf_{r\to+\infty} \frac{\sum_{j=1}^{q} \overline{N}(r, 1/(f - a_j)) + \overline{N}(r,f)}{T(r,f)} \\
&\ge \varliminf_{n\to\infty} \frac{\sum_{j=1}^{q} \overline{N}(r_n, 1/(f - a_j)) + \overline{N}(r_n,f)}{T(r_n,f)} \ge q - 1.
\end{aligned}$$

This yields the inequality

$$\sum_{j=1}^{q} (1 - \Theta(a_j)) + 1 - \Theta(\infty) \ge q - 1,$$

or, equivalently,

$$\sum_{j=1}^{q} \Theta(a_j) + \Theta(\infty) \le 2. \tag{3.20}$$

The above relation is valid for all choices of the sequence of pairwise different complex numbers $a_1, \dots, a_q$; then it follows that the family of nonnegative real numbers

$(\Theta(a))_{a\in\hat{\mathbb{C}}}$ is summable. Though it is well-known that, for such a family, the set of all nonnull entries is countable, we include the proof. According to (3.20), for each $n \in \mathbb{N}$, there are at most $2n-1$ different values of a such that $\Theta(a) > 1/n$. Therefore, the values of a for which $\Theta(a) > 0$ can be arranged in a sequence, in order of decreasing $\Theta(a)$. In fact, first we place the a's satisfying $\Theta(a) = 1$ (note that there are at most two values with this property). Next, we put those a's with $\Theta(a) > 1/2$ but which have not been placed, yet. Continuing in this way, we determine a sequence $(a_n)_{n\geq 0}$, with $a_0 = \infty$. By (3.20), we have

$$\sum_0^q \Theta(a_n) \leq 2 \quad \text{for all } q \in \mathbb{N}.$$

Then, if the sequence (a_n) is infinite, the series $\sum_n \Theta(a_n)$ is convergent and

$$\sum_{n=1}^{\infty} \Theta(a_n) \leq 2.$$

□

Remarks 3.3.5. 1. Suppose that f is nonconstant and meromorphic in $\mathbb{C}$. Then there are at most two values of a such that the equation $f(z) = a$ has no roots.

In fact, assume that the equation $f(z) = a$ has no roots. In this case $N(r, 1/(f-a)) = 0$ for all r and, therefore, $\delta(a) = 1$. This yields $\Theta(a) = 1$ and Nevanlinna's inequality tells us that there may be at most two values of a with this property.

Note that we have $\Theta(a) = 1$ also in the case

$$N(r, 1/(f-a)) = o(T(r,f)) \quad \text{as } r \to +\infty.$$

2. If f is nonconstant and entire, then $\Theta(\infty) = 1$ and Nevanlinna's inequality yields $\sum \Theta(a) \leq 1$, with the sum being extended only to the finite values of a. Therefore, there may be only at most one value of a such that the equation $f(z) = a$ has no roots.

Definition 3.3.6. If f is transcendental meromorphic and $a \in \hat{\mathbb{C}}$, we say that a is a completely branched value if the equation $f(z) - a = 0$ has only a finite number of simple roots.

Proposition 3.3.7. *If f is a transcendental meromorphic (entire, respectively) function, there may be at most four completely branched values (two, respectively).*

Proof. We consider only the case of an entire function. Let q be the number of simple roots of the equation $f(z) - a = 0$. For $r > r_0$, $\overline{n}(r,a) = q + A$, where A denotes the number of multiple roots that the equation has in $|z| < r$, counted once. In the same way, $n(r,a) = q + B$, with B being the number of multiple roots that the equation has in $|z| < r$, counting multiplicities. Note that $A \leq B/2$. Then for $r > r_0$, we have

$$\overline{n}(r,a) = q + A \leq q + \frac{1}{2}(n(r,a) - q) = \frac{1}{2}(q + n(r,a)).$$

Elementary calculus allows us to deduce that

$$\overline{N}(r,a) \leq \frac{1}{2}(q\log r + N(r,a)) + O(1).$$

Now the first fundamental theorem tells us that

$$\overline{N}(r,a) \leq \frac{1}{2}(q\log r + T(r,a)) + O(1).$$

By definition of $\Theta(a)$, via Theorem 3.2.7, it follows that $\Theta(a) \geq 1/2$. In consequence, a transcendental entire function has at most two values of a with this property. □

Theorem 3.3.8 (Clunie theorem). *If f is a transcendental entire function and g is a nonconstant entire function, then*

$$\lim_{r\to+\infty} \frac{T(r,f\circ g)}{T(r,g)} = +\infty.$$

Proof. According to the second fundamental theorem, there is an $a \in \mathbb{C}$ such that the equation $f(z) - a = 0$ has infinitely many roots. Then we may construct a sequence of roots satisfying $|a_{n+1}| > |a_n| + 2$ for all $n \in \mathbb{N}$. For fixed n and $r > 0$, we put $R_0 = \max\{|g(\zeta)| : \zeta \in \overline{\mathbb{D}}_r\}$. As the function $f(z)/(z-a_j)$ is analytic for each $j = 1,\dots,n$, we may choose $M > 0$ such that

$$|f(z) - a_j| < M|z - a_j| \quad \text{for all } |z| < R_0,\ j = 1,\dots,n.$$

Thus we have

$$\log^+ \frac{1}{|f(z) - a|} \geq \log^+ \frac{1}{|z - a_j|} + \log^+(1/M),$$

for all $|z| \leq R_0$ and $j = 1,\dots,n$. Notice that for each z there is at most one $j \leq n$ such that $|z - a_j| < 1$. Then, in fact, we have

$$\log^+ \frac{1}{|f(z) - a|} \geq \sum_{j=1}^{n} \log^+ \frac{1}{|z - a_j|} + \log^+(1/M),$$

for every $|z| < R_0$. Therefore, it follows that

$$m(r, 1/(f\circ g - a)) \geq \sum_{j=1}^{n} m(r, 1/(g - a_j)) + \log^+(1/M).$$

On the other hand, since a root of order p of the equation $g(\zeta) - a_j = 0$ is a root of the equation $f(g(\zeta)) - a = 0$ with order at least p, it follows that

$$N(r, 1/(f\circ g - a)) \geq \sum_{j=1}^{n} N(r, 1/(g - a_j)) + \log^+(1/M).$$

Thus we conclude that

$$T(r, 1/(f \circ g - a)) \geq \sum_{j=1}^{n} T(r, 1/(g - a_j)) + 2\log^+(1/M).$$

Finally, we deduce from the first fundamental theorem that

$$T(r, f \circ g) \geq nT(r, g) + O(1),$$

for every $n \in \mathbb{N}$. □

3.4 Exercises

1. Let f_1 and f_2 be entire functions and denote by $E_j(a)$ the set of points z such that $f_j(z) = a$ $(j = 1, 2)$. Prove that $f_1 \equiv f_2$ or f_1, f_2 are both constant if $E_1(a) = E_2(a)$ for 4 distinct values of a.
2. Consider the functions $f_1(z) = e^z$ and $f_2(z) = e^{-z}$, with $a = 0, \pm 1$, and show that 4 cannot be replaced by 3 in the previous exercise.
3. Let $f(z) = e^z$. Prove that:
   (i) $T(r, f) = r/\pi$.
   (ii) $N(r, a) = r/\pi + O(\log r)$ for $a \in \mathbb{C}$ with $a \neq 0$.
4. Let f be a transcendental entire function. We say that $a \in \mathbb{C}$ has multiplicity at least $m \geq 2$ if all the roots of the equation $f(z) = a$ have multiplicity at least m. Prove that at most four values of this class can exist. Furthermore, if there are four such values, then $m = 2$ for each of them.
5. Let f be a transcendental meromorphic function in the plane. If $f^{(k)}$ denotes the kth derivative of f, prove that

$$\sum_{a \neq \infty} \Theta(a, f^{(k)}) \leq 1 + \frac{1}{k+1}.$$

6. If $f(z) = \tan z$, then $f'(z) = 1/\cos^2 z$. Show that $\Theta(0, f') = 1$ and $\Theta(1, f') = 1/2$. Thus, in this case, the bound $1 + 1/(1 + k)$ in the above exercise is sharp.

Bibliography remarks

Sections 3.1–3.2 are based mainly on [42, Chapter 1] and [66, Chapter VI].

In Section 3.3, we follow [42, Chapter 2].

4 Periodic points

4.1 Attracting and repelling fixed points

We start by showing that a fixed point z^* of an analytic function f is attracting or repelling depending on whether the modulus of the multiplier $f'(z^*)$ is less or greater than 1.

Theorem 4.1.1. *Let f be an entire function and let z^* be a fixed point of f. Then z^* is attracting if and only if $|f'(z^*)| < 1$.*

Proof. (*Necessity*). As $\lim_{n\to\infty} f^n(z) = z^*$ uniformly for $z \in \mathbb{D}_r(z^*)$, there exists an $n_0 \in \mathbb{N}$ such that $|f^n(z) - z^*| < r$ for all $z \in \mathbb{D}_r(z^*)$ and $n \geq n_0$. We consider the function $\phi(z) = (z - z^*)/r$ that conjugates each f^n to $F_n = \phi \circ f^n \circ \phi^{-1}$ for $n \geq n_0$. Then we have

$$F_n(w) = \frac{f^n(rw + z^*) - z^*}{r},$$

with each F_n being defined on $\mathbb{D}$, with values in $\overline{\mathbb{D}}$, and satisfying $F_n(0) = 0$. If we apply Schwarz lemma to each F_n for $n \geq n_0$, we deduce that $|F_n'(0)| \leq 1$ for $n \geq n_0$. We will show that, for some n, the above inequality is strict. By contradiction, suppose that $|F_n'(0)| = 1$ for all $n \geq n_0$. In this case, Schwarz lemma again tells us that there exist a_n with modulus 1 such that $F_n(w) = a_n w$ for all $w \in \mathbb{D}$ and $n \geq n_0$. Therefore

$$|F_n(w)| = |w| = \frac{|f^n(rw + z^*) - z^*|}{r}$$

or, equivalently,

$$|f^n(z) - z^*| = |z - z^*| \quad \text{for all } z \in \mathbb{D}_r(z^*),\ n \geq n_0.$$

For each $z \neq z^*$ in $\mathbb{D}_r(z^*)$, the above condition implies that $\lim_{n\to\infty} f^n(z) \neq z^*$. This is a contradiction, so there exists $n \in \mathbb{N}$ such that $|F_n'(0)| < 1$. Finally, notice that $F_n'(0) = (f^n)'(z^*) = f'(z^*)^n$.

(*Sufficiency*). Choose a $k \in (0, 1)$ such that $|f'(z^*)| < k < 1$. As

$$|f'(z^*)| = \lim_{z\to z^*} \left| \frac{f(z) - f(z^*)}{z - z^*} \right|,$$

there exists an $r > 0$ such that

$$\left| \frac{f(z) - f(z^*)}{z - z^*} \right| < k \quad \text{for all } z \in \mathbb{D}_r(z^*)\ (z \neq z^*).$$

Hence

$$|z - z^*| < r \quad \Longrightarrow \quad |f(z) - z^*| \leq k|z - z^*|.$$

https://doi.org/10.1515/9783111689685-004

Now it is easy to prove by induction on n that

$$|z - z^*| < r \quad \Longrightarrow \quad |f^n(z) - z^*| \le k^n |z - z^*|. \tag{4.1}$$

Since $k^n \to 0$, it follows from (4.1) that z^* is attracting. □

Remark 4.1.2. Let z^* be an attracting fixed point of an entire function f. In the above theorem, the proof of the sufficiency shows that the convergence speed of the orbits starting in a certain neighborhood of z^* depends on the value of $|f'(z^*)|$. In fact, the speed increases as $|f'(z^*)|$ decreases. Also z^* is called superattracting if $f'(z^*) = 0$.

The proof of the necessity part in the next theorem is inspired by [61].

Theorem 4.1.3. *Let f be an analytic function at the fixed point z^*. The fixed point is repelling if and only if $|f'(z^*)| > 1$.*

Proof. (*Necessity*). Obviously, z^* cannot be both repelling and attracting. Then, in view of the above theorem, we know that $|f'(z^*)| \ge 1$. Thus we can take a compact neighborhood V of z^* such that f maps V bianalytically onto $f(V)$. We may choose V small enough such that every orbit starting in V escapes from V. For every $k \ge 1$, the set $V_k = V \cap f^{-1}(V) \cap \cdots \cap f^{-k}(V)$ is a compact neighborhood of z^* consisting of points z for which $f^j(z) \in V$ for all $j = 0, 1, \ldots, k$ (recall that $f^0(z) = z$). Put $V_0 = V$ and notice that the sequence $(V_k)_{k \ge 0}$ is decreasing and $\bigcap_k V_k = \{z^*\}$. By compactness, it follows that $d(V_k) \to 0$ as $k \to \infty$. By the construction, it is easy to show that $f(V_k) = V_{k-1} \cap f(V)$ for all $k \in \mathbb{N}$. Since there is a k_0 such that $V_k \subset f(V)$ for all $k \ge k_0$, we find that $f(V_k) = V_{k-1}$ for $k \ge k_0$. Finally, take $r > 0$ satisfying $\mathbb{D}_r(z^*) \subset V_{k_0}$ and then choose $h > k_0$ such that $V_h \subset \mathbb{D}(z^*, r/2)$. If f^{-1} denotes the inverse of the map $f : V \to f(V)$, put $U = \mathbb{D}_r(z^*)$ and $W = g(U) \subset V_h \subset \mathbb{D}(z^*, r/2)$, where $g = f^{-1} \circ \cdots \circ (h - k_0) \circ \cdots \circ f^{-1}$. As the sets U and W are simply connected, by a standard argument applying Schwarz lemma, we deduce that $|g'(z^*)| < 1$. This concludes the proof since $g'(z^*) = (1/f'(z^*))^{h-k_0}$.

(*Sufficiency*). Choose $k > 1$ such that $|f'(z^*)| > k$. As in the above theorem, there exists a disc $\mathbb{D}_r(z^*)$ such that

$$\left| \frac{f(z) - f(z^*)}{z - z^*} \right| > k \quad \text{for all } z \in \mathbb{D}_r(z^*)\,(z \ne z^*).$$

Equivalently,

$$|f(z) - f(z^*)| \ge k|z - z^*| \quad \text{for all } z \in \mathbb{D}_r(z^*). \tag{4.2}$$

By contradiction, suppose that there exists a $z_0 \in \mathbb{D}_r(z^*)$ $(z_0 \ne z^*)$ such that $f^n(z_0) \in \mathbb{D}_r(z^*)$ for all $n \in \mathbb{N}$. By induction, we can prove that

$$|f^n(z_0) - z^*| \ge k^n |z_0 - z^*|.$$

In fact, since $f^n(z_0) \in \mathbb{D}_r(z^*)$, we can apply (4.2) to obtain

$$|f^{n+1}(z_0) - z^*| = |f(f^n(z_0)) - z^*| \geq k|f^n(z_0) - z^*|.$$

Finally, by the induction hypothesis, it follows that

$$|f^{n+1}(z_0) - z^*| \geq k^{n+1}|z_0 - z^*|.$$

As $k > 1$, we have $\lim_{n\to\infty}(f^n(z_0) - z^*) = \infty$, and this is a contradiction since $f^n(z_0) \in \mathbb{D}_r(z^*)$ for all $n \in \mathbb{N}$. □

Remark 4.1.4. It is important to point out that if z is close to a repelling point z^* (but $z \neq z^*$), the first entries of the orbit of z are further from z^*, but they may come back to the vicinity of z^* (or even to z^* itself) after applying f several times. In fact, we will prove that, when z^* is repelling, $z_n \to z^*$ if and only if $z_n = z^*$ for $n \geq n_0$. By contradiction, suppose that $z_n = f^n(z) \to z^*$ (z^* repelling) and there exists a subsequence $(z_{n_j})_j$ of (z_n) such that $z_{n_j} \neq z^*$ for all $j \in \mathbb{N}$. Since $\lim_{n\to\infty} z_n = z^*$, there is no j_0 satisfying

$$|z_{n_{j+1}} - z^*| \geq |z_{n_j} - z^*| > 0 \quad (\forall j \geq j_0).$$

Thus there exists an infinite number of $j \in \mathbb{N}$ with the property

$$|z_{n_{j+1}} - z^*| < |z_{n_j} - z^*|. \tag{4.3}$$

Now choose a $k > 1$ such that $|f'(z^*)| > k > 1$. As in the proof of the above theorem, take a disc $\mathbb{D}_r(z^*)$ such that

$$|f(z) - f(z^*)| = |f(z) - z^*| > k|z - z^*|,$$

for all $z \in \mathbb{D}_r(z^*)$ ($z \neq z^*$). In particular, if j is sufficiently large, applying $n_{j+1} - n_j$ times the above relation to z_{n_j}, we obtain

$$|z_{n_{j+1}} - z^*| > k^{n_{j+1}-n_j}|z_{n_j} - z^*| \quad \text{(for all } j \geq j_0).$$

This contradicts (4.3).

Examples 4.1.5. 1) The function $f(z) = \mu z(1 - z)$ (logistic complex function) has two finite fixed points:

$$\mu z(1 - z) = z \quad \Longrightarrow \quad z = 0 \text{ or } \mu(1 - z) = 1.$$

Then $f(z)$ has the fixed points $z_1^* = 0$ and $z_2^* = (\mu - 1)/\mu$. Also $f'(z_1^*) = \mu$, so z_1^* is attracting if $|\mu| < 1$ and repelling if $|\mu| > 1$. On the other hand, since $f'(z_2^*) = \mu(1 - 2z_2^*) = 2 - \mu$, it follows that z_2^* is attracting if μ belongs to the open disc $\mathbb{D}(2)$ and repelling for μ in the exterior of the mentioned disc.

2) For the function $f(z) = z^2 + c$, the finite fixed points are $z = (1 \pm \sqrt{1-4c})/2$. If $c = 1/4$, there is only a finite fixed point.
3) The function $f(z) = z + e^z$ has no fixed points.
4) Let $f(z) = z - e^{iz} \sin z$. The fixed points of f are $z^* = k\pi$ $(k \in \mathbb{Z})$. As $f'(z) = 1 - e^{2iz}$, these points are superattracting (this example is due to Fatou).

4.2 The basin of attraction

If z^* is an attracting fixed point of a function f, the *basin of attraction* of z^* is the set

$$\Omega(f, z^*) = \left\{ z_0 \in \mathbb{C} : \lim_{n\to\infty} f^n(z_0) = z^* \right\}.$$

By the definition of an attracting fixed point, there exists a disc $\mathbb{D}_r(z^*)$ satisfying $\overline{\mathbb{D}}_r(z^*) \subset \Omega(f, z^*)$.

Theorem 4.2.1. *If $f : \mathbb{C} \to \mathbb{C}$ is an entire function and z^* is an attracting fixed point of f, then $\Omega(f, z^*)$ is open and completely invariant.*

Proof. (1) Let $a \in \Omega(f, z^*)$ and $r > 0$ be such that $\overline{\mathbb{D}}_{2r}(z^*) \subset \Omega(f, z^*)$. As $\lim_{n\to\infty} f^n(a) = z^*$, there exists a $k \in \mathbb{N}$ such that $|f^k(a) - z^*| < r$. By continuity, there is a $\delta > 0$ such that $|f^k(a) - f^k(z)| < r$ for all $z \in \mathbb{D}_\delta(a)$. Then we have

$$|f^k(z) - z^*| \le |f^k(z) - f^k(a)| + |f^k(a) - z^*| \le 2r,$$

for $|z - a| < \delta$. Hence $f^k(z)$ belongs to the disc $\overline{\mathbb{D}}_{2r}(z^*)$ and this implies that the orbit of $f^k(z)$ converges to z^* for each $z \in \mathbb{D}_\delta(a)$, that is, $\lim_{n\to\infty} f^{k+n}(z) = z^*$ for all $z \in \mathbb{D}_\delta(a)$ and, therefore, $\mathbb{D}_\delta(a) \subset \Omega(f, z^*)$ holds.

(2) If $z_0 \in \Omega(f, z^*)$, then $\lim_{n\to\infty} f^n(z_0) = z^*$. Therefore

$$\lim_{n\to\infty} f^n(f(z_0)) = \lim_{n\to\infty} f^{n+1}(z_0) = z^*,$$

and this proves that $f(z_0) \in \Omega(f, z^*)$. In a similar way, we can prove that $f(z_0) \in \Omega(f, z^*)$ implies $z_0 \in \Omega(f, z^*)$. □

The connected component of $\Omega(f, z^*)$ containing z^* is called the *immediate basin of attraction* of z^* and is denoted by $\Omega_0(f, z^*)$. As $f(\Omega_0(f, z^*))$ is connected and contains the point z^*, it follows that $f(\Omega_0(f, z^*)) \subset \Omega_0(f, z^*)$.

Theorem 4.2.2. *Under the above conditions, $(f^n(z))_n$ converges to z^* uniformly on compact subsets of $\Omega(f, z^*)$.*

Proof. It suffices to prove that the convergence is uniform on a neighborhood of each point of $\Omega(f, z^*)$. By (4.1), we know that

$$|f^n(z) - z^*| \le k^n |z - z^*|$$

for $z \in \mathbb{D}_r(z^*)$, and this implies that $(f^n(z))_n$ converges to z^* uniformly on $\mathbb{D}_r(z^*)$. In the above theorem, we have seen that, for a fixed $a \in \Omega(f, z^*)$, there exist a $k \in \mathbb{N}$ and a disc $\mathbb{D}_\delta(a) \subset \Omega(f, z^*)$ such that $f^k(z)$ belongs to $\mathbb{D}_r(z^*)$ for all $z \in \mathbb{D}_\delta(a)$. As $|f^{n+k}(z) - z^*| = |f^n(f^k(z)) - z^*|$, the uniform convergence of $(f^m)_m$ on $\mathbb{D}_\delta(a)$ follows in an obvious way from the uniform convergence on the disc $\mathbb{D}_r(z^*)$. □

If $f(z) = z \sin z$, the origin is a superattracting fixed point of f. In Figure 4.1, we show the basin of attraction of $z^* = 0$ in the window $[-3, 3] \times [-3, 3]$. If $|f^n(z)| > 10^{-4}$ for every $n \leq 4000$, we decide that the orbit of z does not converge to 0 and draw a black point at z. The other colors depend on the convergence speed of the orbit of z to the origin (this convergence is faster in the red region). The next result will be used throughout the book.

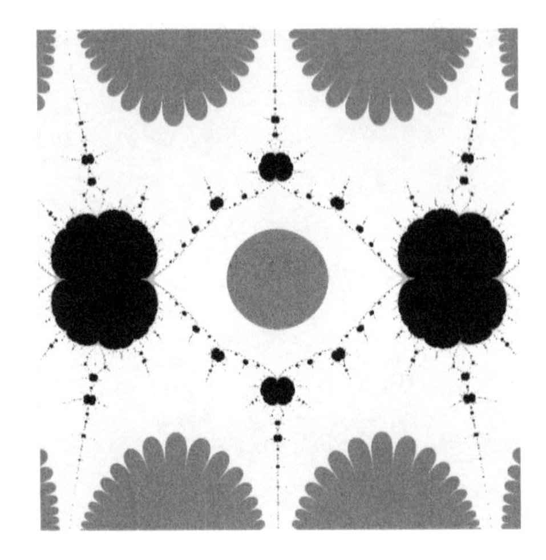

Figure 4.1: Basin of attraction of the origin.

Proposition 4.2.3. *Let f be an entire function and let $\Omega \subset \mathbb{C}$ be a simply connected region, different from $\mathbb{C}$, which is invariant for f. If $z^* \in \Omega$ is an attracting fixed point of f, then Ω is contained in the immediate basin of attraction of z^*.*

The proof is a simple exercise combining Riemann's theorem and Schwarz's lemma (Exercise 9).

4.3 Rationally neutral fixed points

Definition 4.3.1. If z^* is a fixed point of f, we say that z^* is indifferent or neutral if $|f'(z^*)| = 1$. If this is the case, $f'(z^*) = e^{2\pi\theta i}$ where $\theta \in [0, 1]$ is called the rotation number of z^*. When θ is rational, we will say that z^* is rationally indifferent (or parabolic). Otherwise, z^* is called irrationally indifferent.

Irrationally indifferent points will be considered in the next chapter in relation with the problem of the existence of local linearization. In this section, we study the dynamics of a complex function $f(z)$ in a neighborhood of a rationally indifferent point. We will consider two cases of parabolic points:

(1) $f'(z^*) = 1$.
(2) $f'(z^*) = \lambda$ with $\lambda \neq 1$ and there is a positive integer q such that $\lambda^q = 1$.

CASE 1

To simplify, we will suppose that the fixed point is the origin. In a neighborhood of zero, f can be expressed in the form

$$f(z) = z + az^{n+1} + \sum_{k>n+1} a_k z^k, \tag{4.4}$$

where $a \neq 0$ and $n \geq 1$.

Definition 4.3.2 (Attracting and repelling directions). Let u_a be a unit complex number. We will say that u_a is an attracting direction at the origin if $a(u_a)^n$ is real and negative. The unit vector u_r is called a repelling direction at the origin if $a(u_r)^n$ is real and positive.

If the unit vector $u_a \in \mathbb{C}$ is such that $a(u_a)^n$ is a negative real number, then $|a(u_a)^n| = |a|$. Thus the equality $a(u_a)^n = -|a|$ holds. Therefore, u_a has the form

$$u_a = \sqrt[n]{\frac{-|a|}{a}} = \sqrt[n]{\frac{-|a|\bar{a}}{a \cdot \bar{a}}} = \sqrt[n]{\frac{-\bar{a}}{|a|}}.$$

Hence, if α denotes the argument of a, we have

$$u_a = \exp\left(i\left(\frac{\pi - \alpha}{n} + \frac{2\pi}{n}k\right)\right).$$

Then there are n attracting directions, equally spaced, that may be calculated taking $k = 0, 1, \ldots, n-1$ in the above expression (notice that $\arg(-\bar{a}) = \pi - \arg(a)$).

The repelling directions have the same behavior: there are n repelling directions given by

$$u_r = \sqrt[n]{\frac{\bar{a}}{|a|}}$$

and then $u_r = \exp(i(\frac{-\alpha}{n} + \frac{2\pi}{n}k))$, where $k = 0, 1, \ldots, n-1$ (note that $\arg(\bar{a}) = 2\pi - \arg(a)$).

Let us see the reason of this classification. If u_a is an attracting direction for f and $a(u_a)^n = -\alpha$ with $\alpha > 0$, it follows from (4.4) that we can choose an $r > 0$ sufficiently small such that $a(ru_a)^{n+1}$ is a good approximation of $f(ru_a) - ru_a$. That is,

$$f(ru_a) \approx ru_a + a(ru_a)^{n+1} = ru_a + r^{n+1}u_a(a(u_a)^n) = (r - \alpha r^{n+1})u_a = su_a.$$

Setting $z = ru_a$ tells us that $f(z) \approx su_a$, where $s = r - ar^{n+1}$. Note that for all $r < a^{-n}$, we have $0 < s = r - ar^{n+1} < r$. Then for those r, we conclude that $f(z)$ is almost in the "same direction" of $z = ru_a$ and closer to the origin than z.

Definition 4.3.3. (i) We will say that the orbit $z_k = f^k(z_0)$ converges nontrivially to zero if $z_k \to 0$, but $z_k \neq 0$ for all $k \in \mathbb{N}$.

(ii) Let u_a be an attracting direction for f and let $z_0, z_1, \ldots, z_k, \ldots$ be an orbit with $z_k \neq 0$ for all $k \in \mathbb{N}$. We say that (z_k) converges to 0 in the direction u_a if the orbit satisfies

$$z_k \to 0 \quad \text{and} \quad \frac{z_k}{|z_k|} \to u_a.$$

It may be interesting to have a purely local analogue to the notion of a basin of attraction. With this aim, choose open neighborhoods W and W' of zero such that f maps diffeomorphically W onto W' (notice that $f'(0) \neq 0$). Then the inverse function $f^{-1} : W' \to W$ is analytic.

Definition 4.3.4 (Attracting petal). Let u_a be an attracting direction of f at zero. A simply connected region $P \subset W \cap W'$ is called an attracting petal in the direction u_a if the following conditions hold:

(1) $0 \in \partial(P)$ and $f(P) \subset P$,
(2) $\lim_{k\to\infty} f^k(z) = 0$ uniformly in P, and
(3) an orbit $(f^k(z))$ is eventually absorbed by P if and only if it converges to 0 in the direction u_a.

Given $\varepsilon, \delta > 0$ and an attracting direction u_a, $A(u_a, \varepsilon, \delta)$ denotes the angular sector defined by

$$A(u_a, \varepsilon, \delta) = \{z = re^{i\theta}u_a : 0 < r < \delta, |\theta| < (\pi/n) - \varepsilon\}.$$

The next result shows that there exists a suitable attracting petal for every attracting direction. Our proof is based mainly on [61].

Theorem 4.3.5. *Under the above conditions, for each $\varepsilon > 0$ there exists a $\delta > 0$ such that the orbit of every z_0 in the sector $A(u_a, \varepsilon, \delta)$ converges to zero in the direction u_a.*

Proof. (1) Suppose that u_a is the attracting direction given by

$$u_a = \exp\left(i\left(-\frac{\arg(a)}{n} + \frac{\pi}{n}\right)\right).$$

Then u_a bisects the angular region delimited for two consecutive repelling directions

$$-\frac{\arg(a)}{n} < \arg(z) < -\frac{\arg(a)}{n} + \frac{2\pi}{n}. \tag{4.5}$$

We will use the transformation $T(z) = b/z^n$, with $b = -1/(na)$. Its inverse is $T^{-1}(w) = \sqrt[n]{b/w}$. This function is multivalued, and to overcome this problem we will show that the angular region (4.5) is mapped by T to the region of the w-plane given by

$$-\pi < \arg(w) < \pi.$$

In fact, if $|\theta| < \pi/n$, then $z = re^{\theta i}u_a$ belongs to the mentioned region and $w = b/z^n = e^{-n\theta i}/(nr^n|a|)$ (here we have used the equality $a(u_a)^n = -|a|$). Thus $\arg(w) = -n\theta$ and it follows that $\arg(w) \in (-\pi, \pi)$. This allows us to consider the principal argument and in this way to define a uniform inverse of T, namely

$$T^{-1}(w) = \sqrt[n]{\left|\frac{b}{w}\right|}\, e^{i(\pi - \arg(a) - \arg_p(w))/n}, \tag{4.6}$$

where $\arg_p(w)$ denotes the principal argument of w.

Now we will determine the form of the function $g(w)$ conjugate to $f(z)$ through the map $T(z) = b/z^n$. Recall that we assume that the expansion

$$f(z) = z + az^{n+1} + \sum_{k>n+1} a_k z^k = z(1 + az^n + o(z^n)),$$

is convergent in a neighborhood of the origin and $o(z^n)$ verifies

$$\lim_{z\to 0} \frac{o(z^n)}{z^n} = 0.$$

Thus

$$g(w) = T \circ f \circ T^{-1}(w) = \frac{b}{f(z)^n} = \frac{b}{z^n}(1 + az^n + o(z^n))^{-n} = w(1 + az^n + o(z^n))^{-n}.$$

To estimate $(1 + az^n + o(z^n))^{-n}$, we use the binomial series

$$(1+\lambda)^\alpha = 1 + \alpha\lambda + \frac{\alpha(\alpha-1)}{2!}\lambda^2 + \cdots,$$

valid for $|\lambda| < 1$. Applied to our case,

$$(1 + az^n + o(z^n))^{-n} = 1 - n(az^n + o(z^n)) + o(\lambda),$$

with $\lambda = az^n + o(z^n)$. Notice that $\lim_{z\to 0} o(\lambda)/z^n = 0$. Therefore, we have

$$(1 + az^n + o(z^n))^{-n} = 1 - anz^n + o(z^n),$$

for $|z| < r^*$. Finally, since $b = -1/na$, $z^n = b/w$, we obtain

$$g(w) = w\left(1 + \frac{1}{w} + o\left(\frac{b}{w}\right)\right) = w + 1 + w\,o\left(\frac{b}{w}\right) = w + 1 + r(w), \tag{4.7}$$

where $r(w)$ satisfies $\lim_{w\to\infty} r(w) = 0$. As $o(z^n)/z^{n+1} = O(1)$, the slightly more precise statement holds

$$g(w) = w + 1 + O(1/\sqrt[n]{|w|}), \tag{4.8}$$

as $w \to \infty$. In the next chapter, we will need this fact to prove that f is conjugate to a translation $z \to 1 + z$ in an attracting petal.

Thus, given $c \in (0,1)$, there exists an $R_c > 0$ such that

$$|g(w) - w - 1| < c \quad \text{for all } |w| > R_c. \tag{4.9}$$

In particular, if $\mathbb{R}e(w) > R_c$, it follows from (4.9) that

$$\mathbb{R}e(g(w)) = 1 + \mathbb{R}e(w) + \mathbb{R}e(r(w)) > \mathbb{R}e(w) + 1 - c \quad \text{for all } |w| > R_c.$$

Repeating the argument k times, we obtain

$$\mathbb{R}e(g^k(w)) > \mathbb{R}e(w) + k(1-c) \tag{4.10}$$

and, therefore, $\lim_{k\to\infty} g^k(w) = \infty$ uniformly in the half-plane.

Equation (4.10) holds in the half-plane $\mathbb{R}e(w) > R_c$ for all $c \in (0,1)$. However, we will see that it is also true in a larger domain. Consider the two half-lines that are tangent to the circle of radius R_c centered at the origin and with slope $\pm c/\sqrt{1-c^2}$. Let $D(c)$ be the domain on the right, bounded for the curve C determined by those two half-lines and a circle arc, as shown in Figure 4.2.

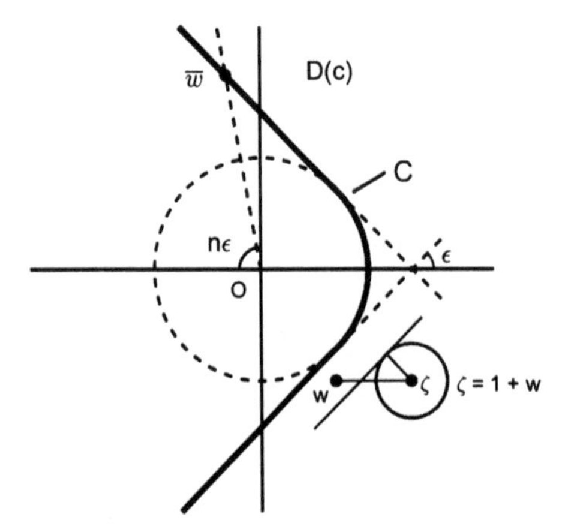

Figure 4.2: The domain $D(c)$.

If $w \in D(c)$ and $\zeta = 1 + w$, then $g(w)$ belongs to the disc $\mathbb{D}_c(\zeta)$ and c is precisely the radius of the circle with center ζ indicated in the figure. Then, $g(D(c)) \subset D(c)$. Also it is obvious that if $w_0 \in D(c)$, its orbit ends up entering $\mathbb{R}e(w) > R_c$ and, therefore, $\lim_{k\to\infty} g^k(w) = \infty$.

Note that for $z \in A(u_a, \varepsilon, \delta)$, we have $\arg(w) \in (-\pi + n\varepsilon, \pi - n\varepsilon)$, where $w = T(z)$.

Given $\varepsilon > 0$, put $c = \sin\varepsilon$ and consider the line with slope $-\tan(n\varepsilon)$ and intersecting the upper tangent at the point $\overline{w}$. Elementary computations show that $|\overline{w}| = R_c/\sin(n-1)\varepsilon$. So, it suffices to choose $\delta = (|b|\sin(n-1)\varepsilon/R_c)^{1/n}$. With this choice of δ, it is easy to prove that T maps the sector $A(u_a, \varepsilon, \delta)$ to a subset of $D(c)$ (in the case $n = 1$, we may put $c = \sin(\varepsilon/2)$).

Claim. $\frac{g^k(w_0)}{k} \to 1$ as $k \to \infty$ for every w_0 in the half-plane $\mathbb{R}e(w) > R_c$.

Given w_0 in the half-plane $\mathbb{R}e(w) > R_c$, put $w_j = g^j(w_0)$. Then (w_j) is the orbit of w_0 and we know that is divergent. For each $k \in \mathbb{N}$, we have

$$\frac{w_k - w_0}{k} = \frac{1}{k}\sum_{j=1}^{k}(w_j - w_{j-1}).$$

By (4.7), each difference $w_j - w_{j-1}$ satisfies

$$w_j - w_{j-1} = g(w_{j-1}) - w_{j-1} = 1 + r(w_{j-1}).$$

Then it follows that

$$\frac{w_k - w_0}{k} = \frac{1}{k}\sum_{j=1}^{k}(1 + r(w_{j-1})) = 1 + \frac{1}{k}\sum_{j=1}^{k} r(w_{j-1}).$$

As $(r(w_{j-1}))$ converges to 0, the sequence of the arithmetic means is also convergent to 0. Thus the above equality implies that

$$\lim_{n\to\infty} \frac{w_k - w_0}{k} = 1. \tag{4.11}$$

Since $w_0/k \to 0$, the proof of the claim is finished.

Now we express this fact in terms of the iterations in the z-plane. We choose z_0 in the sector A and put $w_0 = T(z_0)$. First of all, note that if $z_k = T^k(z_0)$ and $w_k = g^k(w_0)$, then $T(z_k) = w_k$. We know that $w_k/k \to 1$ and will prove that $z_k/|z_k| \to u_a$. We start showing that $z_k/k^{-(1/n)} \to \sqrt[n]{|b|}$. In fact,

$$\frac{z_k}{k^{-(1/n)}} = \frac{T^{-1}(w_k)}{k^{-(1/n)}} = \sqrt[n]{\left|\frac{kb}{w_k}\right|}\, e^{i(\pi - \arg(a) - \arg_p(w_k))/n}.$$

The equality $\lim_{k\to\infty} w_k/k = 1$ implies that $\lim_{k\to\infty}\log(k/|w_k|) = 0$ and $\lim_{k\to\infty}\arg_p(w_k) = 0$. From this we deduce that

$$z_k/k^{-(1/n)} \to \sqrt[n]{|b|}e^{i(\pi - \arg(a))/n}.$$

Finally, we can calculate the limit of the sequence $(z_k/|z_k|)$:

$$\lim_{k\to\infty}\frac{z_k}{|z_k|} = \lim_{k\to\infty}\frac{k^{(1/n)}z_k}{|k^{(1/n)}z_k|} = \frac{\sqrt[n]{|b|}e^{i(\pi-\arg(a))/n}}{\sqrt[n]{|b|}} = e^{i(\pi-\arg(a))/n} = u_a. \qquad \square$$

Remark 4.3.6. The function $f(z) = z + az^{n+1} + o(z^{n+1})$ has an inverse function defined in a disc with center at the origin. It is easy to prove that its Taylor series at 0 is $f^{-1}(z) = z - az^{n+1} + o(z^{n+1})$ and, therefore, the origin is a neutral fixed point for f^{-1} satisfying $(f^{-1})'(0) = 1$. Notice that the attracting directions for f are repelling for f^{-1}, and vice versa. We will give below the definition of a repelling petal as in [61].

Definition 4.3.7. Let v be a repelling direction of f at zero. A simply connected region $P \subset W \cap W'$ is called a repelling petal in the direction v if it is an attracting petal for the inverse map f^{-1} in the direction v.

Proposition 4.3.8. *If P is a repelling petal for f, then for every $z \in P$ there is an $n \in \mathbb{N}$ such that $f^n(z) \notin P$.*

Proof. Suppose that the orbit of z does not escape from P. As P is an attracting petal for f^{-1}, $f^{-m}(f^k(z)) \to 0$ uniformly in $k \in \mathbb{N}$. So, given $\varepsilon > 0$, there is an $m_0 \in \mathbb{N}$ such that $|f^{-m}(f^k(z))| < \varepsilon$ for all $m \geq m_0$ and $k \in \mathbb{N}$. Obviously, this implies that $|f(z)| < \varepsilon$. Since $\varepsilon > 0$ is arbitrary, it follows that $f(z) = 0$. Then $z = 0$ because f is one-to-one in W. $\square$

Therefore, if P is a repelling petal for f and $z \in P$, the orbit of z under f^{-1} converges to zero but the orbit under f escapes from P. This can happen in two different ways: (1) without getting into an attracting petal for f or (2) getting into an attracting petal for f and then its orbit under f converges to 0. The next example shows that case (1) occurs.

If $f(z) = z + z^5$, $z^* = 0$ is a fixed point with $f'(0) = 1$. Then the attracting directions are the 4th roots of -1, $u_k = \pm\sqrt{2}/2 \pm \sqrt{2}/2i$. In Figure 4.3, we see the four attracting petals, while the repelling directions are $\pm 1, \pm i$. If $x > 0$, notice that the sequence $(f^n(x))$

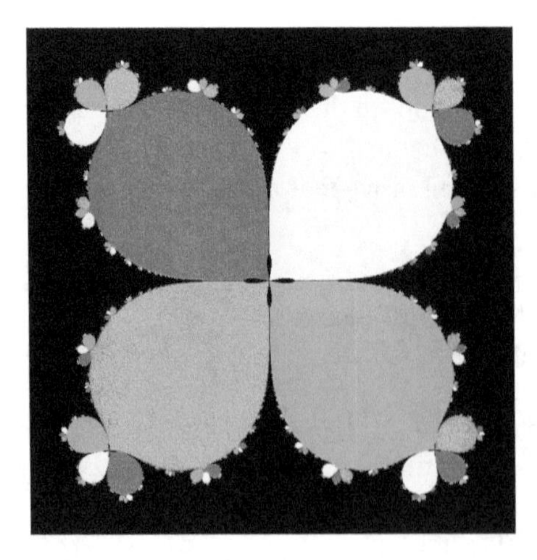

Figure 4.3: The four attracting petals of f at 0.

is increasing and then it converges to a finite fixed point or to infinity. As the origin is the only finite fixed point of f, it follows that $f^n(x) \to \infty$.

In view of Theorem 4.3.5, the following definition arises in a natural way.

Definition 4.3.9. Let z^* be a parabolic fixed point of f with $f'(z^*) = 1$. Given an attracting direction u_j, the parabolic basin of attraction $\mathcal{A}_j = \mathcal{A}(z^*, u_j)$ (in the direction of u_j) is the set of all $z \in \mathbb{C}$ such that the orbit $(f^k(z)) \to z^*$ in the direction u_j.

An alternative definition of $\mathcal{A}_j$ is the following: If P_j is an attracting petal (in the direction u_j), the basin of attraction associated with P_j is the set of all z such that $f^k(z) \in P_j$ for some $k \in \mathbb{N}$. It is obvious that this set is precisely $\mathcal{A}_j$.

The immediate basin $\mathcal{A}_j^0$ may be defined as the connected component of $\mathcal{A}_j$ that contains P_j.

The next proposition lists the basic properties the $\mathcal{A}_j$'s. The proof follows easily applying Theorem 4.3.5.

Proposition 4.3.10. *Under the above conditions, the following statements hold:*
(i) *Each $\mathcal{A}_j$ is open and fully invariant.*
(ii) *The $\mathcal{A}_j$'s are disjoint with the property that an orbit $f^k(z)$ converges to z^* nontrivially if and only if it belongs to one of the $\mathcal{A}_j$'s.*

The flower theorem is also a consequence of Theorem 4.3.5.

Theorem 4.3.11 (The flower theorem of Leau–Fatou). *Under the above conditions, the following statements hold:*
(a) *There exist attracting petals P_i for the n attracting directions and repelling petals P_i' for the n repelling directions such that the union of the $2n$ petals, together with the fixed point, form a neighborhood W_0 of z^**

$$W_0 = \{z^*\} \cup P_1 \cup \cdots \cup P_n \cup P_1' \cup \cdots \cup P_n'.$$

(b) *Two petals intersect if and only if the angle between their central directions is π/n.*

CASE 2

We now focus on the dynamics near a rationally indifferent fixed point z^* with multiplier $\lambda \neq 1$ but equal to a root of unity. Specifically, we suppose that $\lambda^q = 1$, with $q \geq 2$ being a positive integer. By the chain rule, $(f^q)'(z^*) = f'(z^*)^q = 1$. Thus we can apply the flower theorem to f^q in the fixed point z^*.

Example 4.3.12. Let $f(z) = -z + z^4$. In this case, it suffices to obtain f^2:

$$f^2(z) = f(-z + z^4) = -(-z + z^4) + (-z + z^4)^4 = z + 4(-z)^3 z^4 + \cdots = z - 4z^7 + \cdots$$

Then f^2 has 6 attracting petals.

If P is an attracting petal for f^q, then

$$\lim_{n\to\infty} (f^q)^n (z_0) = 0 \quad \text{for all } z_0 \in P.$$

As $f(0) = 0$, from the continuity of f, it follows that

$$\lim_{n\to\infty} (f)^{qn+1} (z_0) = 0 \quad \text{for all } z_0 \in P.$$

Applying f k times, for $k < q$, we obtain

$$\lim_{n\to\infty} (f)^{qn+k} (z_0) = 0 \quad \text{for all } z_0 \in P.$$

This yields $\lim_{n\to\infty} f^n(z_0) = 0$ and, therefore, P is an attracting petal for f. The next claim gives us more information about the number of petals.

Claim. Let $f(z) = \lambda z + az^{N+1} + \cdots$, with λ being a qth root of unity and $a \neq 0$. If n denotes the number of attracting petals of f^q, then q is a divisor of n.

Proof. Note that $\lambda = e^{2\pi p i/q}$, with p and q being positive integers with $p < q$ (we can assume that p and q have no common divisors). Two consecutive attracting directions of f^q form an angle of $2\pi/n$ radians; thus these directions have the form

$$v_0 = e^{i\theta},\ v_1 = e^{i(\theta+2\pi/n)}, \ldots, v_{n-1} = e^{i(\theta+2\pi(n-1)/n)}.$$

Choose a z_0 such that its orbit under f^q converges to 0 in the direction of v_0, that is, $(z_{kq})_k$ satisfies

$$\lim_{k\to\infty} z_{kq} = 0 \quad \text{and} \quad \lim_{k\to\infty} \frac{z_{kq}}{|z_{kq}|} = v_0.$$

Denoting $f^m(z_0)$ by z_m, we show that the orbit of z_1 under f^q, namely $z_1, z_{q+1}, z_{2q+1}, \ldots$, converges to 0 in the direction of λv_0:

$$\lim_{k\to\infty} \frac{z_{kq+1}}{|z_{kq+1}|} = \lim_{k\to\infty} \frac{f(z_{kq})}{|f(z_{kq})|} = \lim_{k\to\infty} \frac{(\frac{f(z_{kq})-f(0)}{z_{kq}-0}) z_{kq}}{|\frac{f(z_{kq})-f(0)}{z_{kq}-0}||z_{kq}|} = \frac{f'(0)}{|f'(0)|} v_0 = \lambda v_0.$$

Hence λv_0 must coincide with some of the v_j ($j \neq 0$). That is,

$$e^{2\pi p i/q} e^{\theta i} = e^{(\theta+(2\pi/n)j)i},$$

and this implies that $(2\pi p)/q - (2\pi j)/n = 2k\pi$. Simplifying, we obtain $p/q - j/n = k$. Since p/q and j/n are positive and less than 1, it follows that necessarily $k = 0$ and, therefore, $pn = qj$. As q and p have no common divisors, it follows that q divides n. □

4.4 Periodic orbits

If z_0 is a p-periodic point of f, then $z_0, f(z_0), f^2(z_0), \ldots, f^{p-1}(z_0)$ are fixed points of f^p. In fact, if we put $z_j = f^j(z_0)$, then

$$f^p(z_j) = f^p(f^j(z_0)) = f^{p+j}(z_0) = f^j(f^p(z_0)) = f^j(z_0) = z_j.$$

Next we will prove that the p points z_j, as fixed points of f^p, have the same multiplier. This allows us to establish the notions of attracting or repelling periodic orbit in a simple way. Let $\{z_0, z_1, z_2, \ldots, z_{p-1}\}$ be a p-cycle for f. By the chain rule,

$$(f^p)'(z_j) = f'(f^{p-1}(z_j))f'(f^{p-2}(z_j)) \cdots f'(f(z_j))f'(z_j) = \prod_{k=0}^{p-1} f'(z_k).$$

Then the p fixed points $z_0, z_1, \ldots, z_{p-1}$ of f^p have the same multiplier.

Theorem 4.4.1. *Let $\{z_0, z_1, \ldots, z_{p-1}\}$ be a p-cycle of f and $K = \prod_{j=0}^{p-1} |f'(z_j)|$. The following statements hold:*
(a) *If $K < 1$, then the p points $z_0, z_1, \ldots, z_{p-1}$ are attracting fixed points of f^p and we will say that the p-cycle is attracting.*
(b) *If $K > 1$, then the points $z_0, z_1, \ldots, z_{p-1}$ are repelling fixed points of f^p and we will say that the p-cycle is repelling.*

Example 4.4.2. We just know that $f(z) = e^z + z$ has no fixed points. However, it has infinitely many periodic points. Next we calculate the 2-periodic points. We must solve the equation $f^2(z) = z$, that is,

$$e^{e^z+z} + e^z + z = z.$$

Simplifying, we see that the 2-periodic points are the solutions of the equation $e^{e^z} + 1 = 0$, namely

$$z_{kh} = \begin{cases} \log((2k+1)\pi) + (2h+1/2)\pi i & \text{if } k \geq 0, \\ \log(|2k+1|\pi) + (2h-1/2)\pi i & \text{if } k < 0. \end{cases}$$

The points z_{kh} are repelling 2-periodic since

$$|f'(z_{kh})| = \sqrt{1 + (\pi(2k+1))^2}.$$

Now we will study the convergence of the orbits to an attracting cycle.

Let $\{z_0^*, z_1^*, \ldots, z_{p-1}^*\}$ be an attracting p-cycle of f. In particular, z_0^* is an attracting fixed point of f^p. Then there is a disc $\mathbb{D}_r(z_0^*)$ such that if $z_0 \in \mathbb{D}_r(z_0^*)$, then

$$\lim_{n\to\infty} (f^p)^n(z_0) = z_0^*.$$

That is,

$$\lim_{n\to\infty} f^{pn}(z_0) = z_0^*.$$

Applying f^k to the above equality, we obtain

$$\lim_{n\to\infty} f^k[f^{pn}(z_0)] = f^k(z_0^*) = z_k^*.$$

Thus for all z_0 in the disc $\mathbb{D}_r(z_0^*)$, one gets

$$\lim_{n\to\infty} f^{pn+k}(z_0) = z_k^* \quad \text{for } k = 0, 1, 2, \ldots, p-1.$$

Then the orbit $(f^n(z_0))_n$ has p oscillation limits since each subsequence $(f^{pn+k}(z_0))_n$ converges to z_k^*.

Obviously, for each point z_k^* of the p-cycle, there exists a disc $\mathbb{D}_r(z_k^*)$ with this same property.

In order to establish the definition of a basin of attraction of a p-cycle, consider an attracting p-cycle $\{z_0^*, z_1^*, \ldots, z_{p-1}^*\}$ for f. If Ω_k denotes the basin of attraction of z_k^* for f^p, then $\Omega = \Omega_0 \cup \Omega_1 \cup \cdots \cup \Omega_{p-1}$ is called the *basin of attraction of the p-cycle*. This Ω has the following property:

$$z_0 \in \Omega_k \quad \Longrightarrow \quad f(z_0) \in \Omega_{k+1}.$$

In fact, $z_0 \in \Omega_k$ means that $\lim_{n\to\infty}(f^p)^n(z_0) = z_k^*$. Since f is continuous, applying f to both sides of the equality, we obtain

$$f\Big(\lim_{n\to\infty}(f^p)^n(z_0)\Big) = \lim_{n\to\infty}(f^p)^n(f(z_0)) = f(z_k^*) = z_{k+1}^*,$$

which proves that $f(z_0)$ belongs to Ω_{k+1}. Again, the immediate basin of attraction of an attracting cycle $\{z_0^*, z_1^*, \ldots, z_{p-1}^*\}$ is the union of the immediate basins of attraction of each z_k^* (as fixed points of f^p).

In Theorem 10.3.5, it is proved that $Q_c(z) = z^2 + c$ has an attracting 2-cycle $\{z_0, z_1\}$ for $|c+1| < 1/4$, where z_0 and z_1 are the roots of $z^2 + z + c + 1$. In Figure 4.4, we see the basin of attraction of the attracting 2-cycle $\{z_0, z_1\}$ of the function $f(z) = z^2 - 1 + 0.2i$, where $z_0 = 0.0339 - 0.1873i$ and $z_1 = -1.0339 + 0.1873i$ (in the window $[-1.5, 1.5] \times [-1.5, 1.5]$). The yellow region is the basin of z_1 as an attracting fixed point of f^2.

Obviously, a rational function has finitely many p-periodic points for every $p \in \mathbb{N}$. This is not the case for transcendental entire functions. The following result was proved by Baker [4] in 1960, applying the second theorem of Nevanlinna.

Theorem 4.4.3. *If f is a transcendental entire function, then, for every $n \in \mathbb{N}$, f has infinitely many periodic points of period n, except for at most one integer n.*

Proof. Suppose that k is the smallest positive integer such that f has a finite number of k-periodic points. We shall prove that f has infinitely many n-periodic points for all

$n > k$. Arguing by contradiction, assume that there is an $n > k$ such that f has only a finite number q of n-periodic points. We consider the function $h(z) = (f^n(z)-z)/(f^k(z)-z)$ and apply the second fundamental theorem of Nevanlinna to the values 0, 1, and ∞:

$$T(r,h) \leq \overline{N}(r,1/(h-1)) + \overline{N}(r,1/h) + \overline{N}(r,h) + o(T(r,h)). \tag{4.12}$$

By hypothesis, if p is the number of k-periodic points of f, then

$$\overline{N}(r,1/h) \leq q\log r + O(1) \quad \text{and} \quad \overline{N}(r,h) \leq p\log r + O(1),$$

for r sufficiently large. Note that if z_0 is a root of the equation $h(z) = 1$, then $w_0 = f^k(z_0)$ is a root of $f^{n-k}(w) - w = 0$. This yields

$$\overline{N}(r,1/h-1) \leq \overline{N}(r,1/(f^{n-k}(z)-z)).$$

Using the property of the sum for the characteristic function, we deduce

$$\overline{N}(r,1/(f^{n-k}(z)-z)) \leq T(r,f^{n-k}) + \log r + O(1).$$

With these relations, (4.12) adopts the final form

$$T(r,h) \leq T(r,f^{n-k}) + (p+q+1)\log r + o(T(r,h)).$$

On the other hand, by Clunie's theorem 3.3.8, it follows that $T(r,h)$ has the same growth as $T(r,f^n)$. Therefore, $T(r,f^{n-k})/T(r,h) \to 0$ as $r \to +\infty$. Finally, dividing both sides of the above inequality by $T(r,h)$ and letting $r \to +\infty$, we get $1 \leq 0$. □

Corollary 4.4.4. *If f is a transcendental entire function and $n \geq 2$, then f^n has infinitely many fixed points.*

Example 4.1.5(3) shows that for $n = 1$ the above result does not hold.

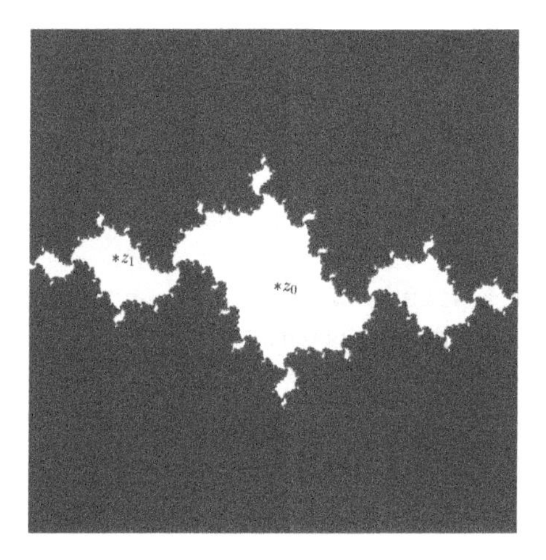

Figure 4.4: Basin of attraction of the 2-cycle.

4.5 Exercises

1. Let $f(z) = z + \lambda \sin z$ with $\lambda \neq 1$. Show that every solution of the equation $\sin z = (2k+1)\pi/\lambda$ is a repelling 2-periodic point of f for $k \in \mathbb{Z}$ whenever $2k+1 > \sqrt{2 + |\lambda|^2}/\pi$.
2. Let $f(z) = \lambda z e^z$ for $|\lambda| < 1$. Determine the best $r > 0$ such that $\mathbb{D}_r$ is contained in the attractive basin of the origin.
3. Let $f(z, w)$ be a function defined on $\mathbb{C} \times \Omega$, with $\Omega \subset \mathbb{C}$ being a region. Assume that (a) there is a $w_0 \in \Omega$ such that $f(z, w_0)$ has a repelling fixed point z_0, and (b) $f(z, w)$ is analytic in $\mathbb{C} \times \Omega$. Prove that, given $\varepsilon > 0$, there exists a $\delta > 0$ such that $f(z, w)$ has a repelling fixed point $z(w)$ satisfying $|z(w) - z_0| < \varepsilon$, provided $|w - w_0| < \delta$.
4. Let $\Omega \subset \mathbb{C}$ be a region and let f be an analytic function from Ω onto itself. If the complement of Ω contains at least two points and f has an attracting fixed point z^* in Ω, then (f^n) is uniformly convergent to z^* on compact subsets of Ω.
5. Let f be an entire function and let z^* be a repelling periodic point for f. Prove directly, using the definitions, that (f^n) is not normal in a neighborhood of z^*.
6. Let $f(z) = \lambda z + az^2$, where $a \neq 0$ and $\lambda^3 = 1$, but $\lambda \neq 1$. Determine the number of petals around the origin.
7. Prove that the immediate basin of attraction of an attracting periodic point is simply connected.
8. Let $f(z) = \log(1 + z)$ for $z \in U = \{z \in \mathbb{C} : \mathrm{Re}(z) > 0\}$, where log denotes the principal branch of the logarithm. Show that $f(U) \subset U$ and $f^n(z) \to 0$ locally uniformly on U.
9. Let f be an entire function and let $U \subset \mathbb{C}$ be a simply connected region different from $\mathbb{C}$. If U is invariant under f and $z_0 \in U$ is an attracting fixed point of f, prove that U is contained in the immediate attractive basin of z_0.
10. (i) Show that every polynomial $P(z) = a_n z^n + a_{n-1} z^{n-1} + \cdots + a_1 z + a_0$, with $a_n \neq 0$, is conjugate to a polynomial of the form $Q(z) = z^n + b_{n-2} z^{n-2} + \cdots + b_1 z + b_0$.
    (ii) Determine all values of the real numbers a and b such that $P(z) = z^3 + az + b$ has a fixed point with multiplier 1.
    (iii) Describe, in this case, the dynamics of $P(z)$ near the origin.
11. Determine (up to conjugation) all polynomials of degree 3 having exactly one attracting fixed point.

Bibliography remarks

In Section 4.1, the proof of the necessity part in Theorem 4.1.3 is inspired by [61, Section 8].

Section 4.3 is based on [61, Section 10]. In Proposition 4.3.8, we prove an interesting property of repelling petals which is not considered in most books.

In the final section, we follow [4].

5 Local conjugacy

In this chapter, we will study the so-called problem of the analytic conjugation in a neighborhood of a fixed point. To simplify, we will suppose that the fixed point is the origin. Then, the Taylor series of f has the form

$$f(z) = \lambda z + a_2 z^2 + a_3 z^3 + \cdots,$$

being convergent for $|z| < r$.

Obviously, we wish the conjugate function g to be as simple as possible. In the first section, we consider the case of a fixed point whose multiplier λ satisfies $|\lambda| \neq 0, 1$ and it is proved that there is a conformal map that conjugates f to $g(z) = \lambda z$ in a neighborhood of the origin. When the fixed point z^* is superattracting, f is conjugate to $g(z) = z^q$, with q being the multiplicity of z^* as a zero of $f(z) - z$.

The first result of this type is due to Koenigs (1884), who proved the existence of linearization in the case of an attracting fixed point [54]. In 1904, Böttcher solved the superattracting case [16]. The parabolic case, when the multiplier is a root of unity, was investigated by Leau. In his study of this hard problem, he proved the famous flower theorem.

5.1 Local linearization for an attracting point

We need the following lemma.

Lemma 5.1.1. *Given $a \in (0,1)$, there exists a $c \in (0,1)$ such that $c^2 < a < c$.*

Proof. We will choose $\epsilon > 0$ so that $a+\epsilon$ is the required c. For this, we take an ϵ satisfying $(a+\epsilon)^2 < a$ or, equivalently, $g(\epsilon) = \epsilon^2+2a\epsilon+a^2-a < 0$. The zeros of $g(\epsilon)$ are $\epsilon_1 = -a+\sqrt{a} > 0$ and $\epsilon_2 = -a - \sqrt{a} < 0$ and, therefore, $g(\epsilon) = (\epsilon - \epsilon_1)(\epsilon - \epsilon_2)$. Notice that for $\epsilon \in (0, \epsilon_1)$ we have $g(\epsilon) < 0$. So, it suffices to show that $c = a + \epsilon < 1$: $a + \epsilon < a + \epsilon_1 = a + \sqrt{a} - a = \sqrt{a} < 1$. □

Theorem 5.1.2 (Koenigs, 1884). *Let $f(z) = \lambda z + \sum_{n=2}^{\infty} a_n z^n$ be convergent for $|z| < r$. If $|\lambda| \neq 1, 0$, then there exist a neighborhood U of 0 and a conformal map $\phi : U \to \mathbb{C}$ such that $\phi \circ f \circ \phi^{-1}(w) = \lambda w$. This ϕ is unique up to multiplication by a nonzero constant.*

Proof. 1) *Proof of the existence for the case* $0 < |\lambda| < 1$. According to Lemma 5.1.1, setting $a = |\lambda|$, we can choose a $c > 0$ such that $c^2 < |\lambda| < c < 1$. As $|f'(0)| = |\lambda| < c < 1$, there is a $\delta > 0$ such that

$$|f(z)| \le c|z| \quad \text{for all } z \in \overline{D}_\delta. \tag{5.1}$$

On the other hand, $f(z) - \lambda z = \sum_{n\ge2} a_n z^n = z^2(\sum_{n\ge2} a_n z^{n-2})$. Then we have

$$|f(z) - \lambda z| \le k|z|^2 \quad (z \in \overline{D}_\delta), \tag{5.2}$$

https://doi.org/10.1515/9783111689685-005

where $k = \max\{|\sum_{n\geq 2} a_n z^{n-2}| : z \in \overline{D}_\delta\}$. By induction on $n \in \mathbb{N}$, it follows from (5.1) that

$$|f^n(z) - 0| \leq c^n|z| \leq c^n\delta < \delta \quad (z \in \overline{D}_\delta). \tag{5.3}$$

As $c \in (0,1)$, (5.3) implies that $(f^n(z))_n \to 0$ uniformly on $\overline{D}_\delta$. Now we estimate the difference $|f^{n+1}(z) - \lambda f^n(z)|$:

$$|f^{n+1}(z) - \lambda f^n(z)| = |f(f^n(z)) - \lambda f^n(z)| \leq k|f^n(z)|^2$$

(in the last step we have applied (5.2) to $w = f^n(z) \in \overline{D}_\delta$). From the above inequality and (5.3), we obtain

$$|f^{n+1}(z) - \lambda f^n(z)| \leq k|f^n(z)|^2 \leq kc^{2n}|z|^2 \quad (z \in \overline{D}_\delta), \tag{5.4}$$

for all $z \in \overline{D}_\delta$. Now we define $\phi_n(z) = f^n(z)/\lambda^n$ $(z \in \overline{D}_\delta)$ and prove that $(\phi_n(z))_n$ converges uniformly to an analytic function in D_δ. Indeed, for $z \in \overline{D}_\delta$, we have

$$|\phi_{n+1}(z) - \phi_n(z)| = \frac{1}{|\lambda|^{n+1}}|f^{n+1}(z) - \lambda f^n(z)| \leq \frac{k}{|\lambda|}\left(\frac{c^2}{|\lambda|}\right)^n |z|^2. \tag{5.5}$$

By a standard argument using (5.5), we can prove that $(\phi_n(z))_n$ is uniformly convergent on $\overline{D}_\delta$ to an analytic function $\phi(z)$ in D_δ. Now we will show that $\phi(z)$ satisfies the required conditions:

(a) First, $\phi(f(z)) = \lambda\phi(z)$ for $z \in D_\delta$:

$$\phi(f(z)) = \lim_{n\to\infty} \frac{f^n(f(z))}{\lambda^n} = \lim_{n\to\infty} \frac{f^{n+1}(z)}{\lambda^n} = \lambda\phi(z).$$

In particular, this yields $\phi(0) = 0$.

(b) Second, there exists an $M > 0$ such that $|\phi(z) - z| < M|z|^2$ for $z \in \overline{D}_\delta$:

$$\begin{aligned}|\phi_n(z) - z| &\leq |\phi_n(z) - \phi_{n-1}(z)| + \cdots + |\phi_1(z) - z| \\ &\leq \sum_{i=2}^{n} |\phi_i(z) - \phi_{i-1}(z)| + \frac{1}{|\lambda|}|f(z) - \lambda z|.\end{aligned}$$

This, together with (5.2) and (5.5), produces

$$|\phi_n(z) - z| \leq |z|^2 \frac{k}{|\lambda|} \sum_{i=2}^{n} \left(\frac{c^2}{|\lambda|}\right)^{i-1} + |z|^2 \frac{k}{|\lambda|} \leq M|z|^2,$$

for $z \in D_\delta$. Passing to the limit as $n \to \infty$, we obtain the inequality in (b).

(c) Last, $\phi'(0) = 1$. In fact,

$$\phi'(0) = \lim_{z\to 0} \frac{\phi(z) - \phi(0)}{z - 0} = \lim_{z\to 0} \frac{\phi(z)}{z}.$$

Finally, applying (b), we find

$$\left|\frac{\phi(z)}{z} - 1\right| \le M|z| \quad (0 < |z| < \delta),$$

which implies that $\phi'(0) = 1$ (recall that $\phi(0) = 0$). By (c), ϕ is conformal in a neighborhood of 0.

2) *The existence in the case* $|\lambda| > 1$. The inverse function f^{-1}, which is defined in a neighborhood of the origin, verifies all the conditions in case (1): the origin is a fixed point of f^{-1} and $|(f^{-1})'(0)| = 1/|\lambda| < 1$. Then there exists a conformal map $\phi : U \to \mathbb{C}$ satisfying $\phi(f^{-1}(w)) = (1/\lambda)\phi(w)$, for all $w \in U$. Thus $\lambda\phi(f^{-1}(w)) = \phi(w)$ and, applying this equality to $w = f(z)$, we obtain $\lambda\phi(z) = \phi(f(z))$.

3) *Uniqueness.* Suppose that there exist ϕ and ψ verifying:

$$\phi \circ f \circ \phi^{-1} = L = \psi \circ f \circ \psi^{-1},$$

where $L(w) = \lambda w$. Then $\psi \circ \phi^{-1}$ and L are permutable maps. In fact, the equality $\phi^{-1} \circ L = f \circ \phi^{-1}$ implies that

$$\psi \circ \phi^{-1} \circ L = \psi \circ f \circ \phi^{-1}. \tag{5.6}$$

On the other hand, the equality $\psi \circ f = L \circ \psi$ and (5.6) yield

$$\psi \circ \phi^{-1} \circ L = L \circ \psi \circ \phi^{-1}.$$

Since $L(w) = \lambda w$, from the latter equality we deduce that

$$\psi \circ \phi^{-1}(\lambda w) = \lambda \psi \circ \phi^{-1}(w), \tag{5.7}$$

for w in a neighborhood of the origin. As $\psi \circ \phi^{-1}(0) = 0$, we can express $\psi \circ \phi^{-1}$ in the form

$$\psi \circ \phi^{-1}(w) = \sum_{n=1}^{\infty} b_n w^n.$$

Finally, (5.7) implies that

$$\sum_{n=1}^{\infty} b_n (\lambda w)^n = \lambda \sum_{n=1}^{\infty} b_n w^n,$$

for w in a neighborhood of the origin. Since the respective coefficients are equal, we conclude that $b_n \lambda^n = \lambda b_n$, for each $n \in \mathbb{N}$. If $b_n \neq 0$ for some $n \ge 2$, then $\lambda^n = \lambda$, which is not possible because $|\lambda| \neq 0, 1$. Therefore, $b_n = 0$ for $n \ge 2$ and then $\psi \circ \phi^{-1}(w) = b_1 w$. In consequence, we have $\phi = \psi / b_1$. □

In Chapter 6, we will need the following result. According to the above theorem, if z^* is an attracting fixed point of f with multiplier λ satisfying $0 \neq |\lambda| < 1$, then there exist a neighborhood W of z^* and a conformal map $\phi : W \to \mathbb{D}_\delta$ such that $\phi(z^*) = 0$ and

$$\phi(f(z)) = \lambda\phi(z) \quad \text{for } z \in W. \tag{5.8}$$

Corollary 5.1.3. *Under the above conditions, $f^n(z) \neq z^*$ for every $z \in W$ and $n \in \mathbb{N}$.*

Proof. If $z \in W$ and $f^n(z) = z^*$, then it follows from (5.8) that $\phi(f^n(z)) = \lambda^n\phi(z)$. Then $0 = \phi(z^*) = \lambda^n\phi(z)$ and, in consequence, $\phi(z) = 0$, which contradicts the fact that ϕ is univalent in W. □

It follows easily from the above corollary that every $z \in W$ has an infinite orbit.

Remark 5.1.4. The function $\phi(z)$ may be extended to the whole attractive basin $\Omega(f, z^*)$ preserving (5.8). In fact, if $z_0 \in \Omega(f, z^*)$, then $\lim_{n\to\infty} f^n(z_0) = z^*$, so there exists an $m \in \mathbb{N}$ such that $f^m(z_0) \in \mathbb{D}_\delta(z^*)$ and we can define $\hat{\phi}(z_0) = \lambda^{-m}\phi(f^m(z_0))$. In this way, we get a function $\hat{\phi}$ defined on $\Omega(f, z^*)$. Using (5.8), we can prove easily that this definition is independent of the choice of m. To prove that $\hat{\phi}$ is analytic, notice that, for a fixed $z_0 \in \Omega(f, z^*)$, $f^n(z)$ converges uniformly on a neighborhood V of z_0. Then there exists an m such that $f^m(z) \in D_\delta$ for all $z \in V$ and we can define $\hat{\phi}(z) = \lambda^{-m}\phi(f^m(z))$.

From the definition of $\phi(z)$ for $z \in D_\delta$, it follows that

$$\hat{\phi}(z) = \lambda^{-m} \lim_{n\to\infty} \phi_n(f^m(z)) = \lim_{n\to\infty} \frac{f^{m+n}(z)}{\lambda^{m+n}}. \tag{5.9}$$

Now it is easy to prove that $\hat{\phi}$ verifies the identity $\hat{\phi}(f(z)) = \lambda\hat{\phi}(z)$ for all $z \in \Omega(f, z^*)$.

If z^* is a fixed point of f and $\lambda \neq 0$ is the multiplier, the equation

$$\phi(f(z)) = \lambda\phi(z) \tag{5.10}$$

is called the *Schröder equation* for f. We say that f is linearizable at z^* if there exists a univalent function ϕ defined in a neighborhood U of z^* satisfying the Schröder equation and the condition $\phi(z^*) = 0$. The case $|\lambda| = 1$ is very complicated. We gather several basic facts in Section 5.3.

5.2 Local conjugacy for a superattracting point

Let f be a function so that the origin is a superattracting fixed point. We have, in a neighborhood of the origin,

$$f(z) = b_0 z^q + b_1 z^{q+1} + \cdots,$$

with $b_0 \neq 0$ and $q \geq 2$. In this section we follow mainly the ideas of [13].

We will need the following lemma.

Lemma 5.2.1. *Let $n \geq 2$ and $D = \{w : |w - 1| < (1/2)\}$. If $z^{1/n}$ denotes the principal branch of the nth root defined on D, then*

$$|z^{1/n} - 1| < 2|z - 1|/n < 1/n.$$

Proof. For each $z \in D$, let γ be the segment joining 1 to z. Then

$$\int_\gamma \phi'(z)dz = z^{1/n} - 1,$$

where $\phi(z) = z^{1/n}$. Hence

$$|z^{1/n} - 1| \leq \int_\gamma |\phi'(z)||dz| \leq |z - 1| \max\{|\phi'(w)| : w \in \gamma\}.$$

As $\phi'(z) = z^{1/n}/(nz)$ for $z \in D$, we have

$$|\phi'(z)| = \frac{1}{n|z|^{1-(1/n)}} \leq \frac{2^{1-(1/n)}}{n} < \frac{2}{n}.$$

Finally,

$$|z^{1/n} - 1| \leq \frac{2|z - 1|}{n} < \frac{1}{n}.$$

□

Theorem 5.2.2. *Under the above conditions, there exists a unique function g defined and analytic in a neighborhood of the origin such that*
(a) $g(0) = 0$ *and* $g'(0) = 1$.
(b) $g \circ f \circ g^{-1}(z) = z^q$.

Proof. Conjugating with the function $\phi(z) = \lambda z$, where $\lambda^{q-1} = b_0$, we can suppose that $b_0 = 1$. Then the Taylor series at the origin adopts the form

$$f(z) = z^q + b_1 z^{q+1} + \cdots = z^q(1 + h(z)), \tag{5.11}$$

which is convergent for $|z| \leq R$. Since $h(z) = z(b_1 + b_2 z + \cdots)$, it follows that

$$|h(z)| \leq K|z|, \quad \text{for } |z| \leq R, \tag{5.12}$$

where $K = \max\{|h(z)/z| : |z| \leq R\}$. Choosing $\delta < \min\{1/4, R, 1/(2k)\}$, we have for $z \in \mathbb{D}_\delta$,

$$|f(z)| = |z|^q|1 + h(z)| \leq |z|^q(1 + \delta K) \leq |z|\delta^{q-1}(1 + \delta K) < \frac{|z|}{2}.$$

Thus $f(\mathbb{D}_\delta) \subset \mathbb{D}_\delta$ and it is easy to deduce that

$$|f^n(z)| \leq \frac{|z|}{2^n} \quad (z \in \mathbb{D}_\delta). \tag{5.13}$$

For $z \in \mathbb{D}_\delta$ and $n \geq 0$, by (5.12) and (5.13), we get

$$|h(f^n(z))| \leq K|f^n(z)| \leq K\frac{|z|}{2^n} < \frac{1}{2},$$

which shows that $1+h(f^n(z))$ belongs to the set D of the above lemma and we can consider the q^{n+1}th root of $1 + h(f^n(z))$ that will be denoted by $h_{n+1}(z)$ for $z \in \mathbb{D}_\delta$. By Lemma 5.2.1, we know that

$$|h_{n+1}(z) - 1| < \frac{1}{q^{n+1}} \quad (z \in \mathbb{D}_\delta). \tag{5.14}$$

Now consider the infinite product

$$g(z) = z\prod_{n=1}^{\infty} h_n(z),$$

which is uniformly convergent on $\mathbb{D}_\delta$ by (5.14). It is obvious that $g(0) = 0$ and $g'(0) = 1$. To prove (b), we will show that the following relation holds:

$$g_n(f(z)) = g_{n+1}(z)^q \quad \text{for all } n \in \mathbb{N},\ z \in \mathbb{D}_\delta, \tag{5.15}$$

where $g_n(z) = zh_1(z)\cdots h_n(z)$. For this, we need the equality

$$h_{n+1}(f(z)) = h_{n+2}(z)^q \quad \text{for all } n \in \mathbb{N},\ z \in \mathbb{D}_\delta. \tag{5.16}$$

In fact, we have

$$[h_{n+1}(f(z))]^{q^{n+1}} = 1 + h(f^{n+1}(z)) = [h_{n+2}(z)^q]^{q^{n+1}},$$

which implies (5.16). Finally, we prove (5.15) as follows:

$$g_n(f(z)) = f(z)h_1(f(z))\cdots h_n(f(z)) = f(z)[h_2(z)\cdots h_{n+1}(z)]^q = g_{n+1}(z)^q\frac{f(z)}{(zh_1(z))^q}.$$

From (5.11), having in mind the definition of $h_1(z)$, it follows that

$$f(z)/(zh_1(z))^q = 1, \quad \text{for all } z \in \mathbb{D}_\delta.$$

This yields (5.15) and, passing to the limit as $n \to \infty$, we obtain (b). □

5.3 Local conjugacy for a parabolic fixed point

We need the following useful variant of the well-known Schwarz's lemma.

Lemma 5.3.1. *If f is an analytic function mapping $\mathbb{D}_r(z_0)$ into some disc of radius s, then*

$$|f'(z_0)| \le s/r.$$

Proof. If w_0 is the center of the disc of radius s containing $f(\mathbb{D}_r(z_0))$, set $F(z) = f(z + z_0) - w_0$. Then F maps $\mathbb{D}_r$ into $\mathbb{D}_s$ and, applying the Cauchy integral formula, we have

$$f'(z_0) = F'(0) = \frac{1}{2\pi i} \oint_{|z|=r_1} \frac{F(z)\,dz}{z^2},$$

for all $0 < r_1 < r$. By a straightforward argument, the proof concludes. □

Throughout this section, f is an analytic function in a neighborhood of a fixed point z^* with multiplier $f'(z^*) = 1$. To simplify, we suppose that $z^* = 0$. Then in a neighborhood of the origin, f has the expansion

$$f(z) = z + az^{n+1} + \sum_{k>n+1} a_k z^k, \tag{5.17}$$

with $a \neq 0$ and $n \geq 1$. We will use the notations of the proof of Theorem 4.3.5. Recall that the map $T(z) = b/z^n$ maps the angular region

$$-\frac{\arg(a)}{n} < \arg(z) < -\frac{\arg(a)}{n} + \frac{2\pi}{n}, \tag{5.18}$$

delimited by two consecutive repelling directions, onto the region of the w-plane given by

$$-\pi < \arg(w) < \pi,$$

where $b = -1/(na)$. This allows us to consider the principal argument and in this way we can define a uniform inverse of T.

If g denotes the conjugate of $f(z)$, through the map $T(z) = b/z^n$, we showed that

$$g(w) = w + 1 + O(1/\sqrt[n]{|w|}), \tag{5.19}$$

for $w \to \infty$. Then we can choose positive constants R_0 and C such that

$$|g(w) - 1 - w| < C/\sqrt[n]{|w|}, \tag{5.20}$$

for $|w| > R_0/2$. If we take an $R > R_0$ such that

$$|g(w) - 1 - w| < \frac{1}{2}, \tag{5.21}$$

for $|w| > R$, then the half-plane $\mathbb{R}e(w) > R$ is invariant under g. Indeed, if $\mathbb{R}e(w) > R$, then

$$\mathbb{R}e(g(w)) = 1 + \mathbb{R}e(w) + \mathbb{R}e(r(w)) > 1 + \mathbb{R}e(w) - 1/2 = \mathbb{R}e(w) + 1/2,$$

where $r(w) = g(w) - 1 - w$. By induction, we obtain

$$\mathbb{R}e(g^k(w)) > \mathbb{R}e(w) + \frac{k}{2}, \tag{5.22}$$

for all w in the half-plane $H_R = \{w : \mathbb{R}e(w) > R\}$ and $k \geq 1$.

The main result of this section, due to Leau and Fatou, is Theorem 5.3.6. Our proof is inspired by [61]. We will need the following technical lemmas.

Lemma 5.3.2. *Under the above conditions, there exists a positive constant C' such that*

$$\left|\frac{g^k(w) - g^k(\hat{w})}{g^{k-1}(w) - g^{k-1}(\hat{w})} - 1\right| \leq \frac{C'}{k^{1+1/n}}, \tag{5.23}$$

for all w and $\hat{w}$ in the half-plane H_R.

Proof. For all $|w_0| > R_0$, the function $w \to g(w) - 1 - w$ maps the disc of radius $|w_0|/2$ centered at w_0 into a disc of radius $C\sqrt[n]{2/|w_0|}$. According to Lemma 5.3.1, we have the following estimate for the first derivative of g:

$$|g'(w) - 1| < C\left(\frac{2}{|w|}\right)^{1+1/n}, \tag{5.24}$$

for all $|w| > R_0$.

Finally, since $g(w) - g(\hat{w}) = \int_{\hat{w}}^{w} g'(\zeta)\, d\zeta$ for any two points w and $\hat{w}$ in the half-plane H_R, it follows from (5.24) that

$$\left|\frac{g(w) - g(\hat{w})}{w - \hat{w}} - 1\right| \leq C\left(\frac{2}{\psi}\right)^{1+1/n},$$

where $\psi \geq R$ denotes the smallest of the two numbers $\mathbb{R}e(w)$ and $\mathbb{R}e(\hat{w})$. By (5.22), we know that $\mathbb{R}e(g^k(w))$ and $\mathbb{R}e(g^k(\hat{w}))$ increase at least linearly with k. Then, in particular, the above inequality yields

$$\left|\frac{g^k(w) - g^k(\hat{w})}{g^{k-1}(w) - g^{k-1}(\hat{w})} - 1\right| \leq \frac{C'}{k^{1+1/n}}, \tag{5.25}$$

with $C' > 0$ being a suitable constant. □

Remark 5.3.3. Applying (5.24), it is easy to deduce that g is univalent in the half-plane $\mathbb{R}e(w) > R_0$ (use the equality $g(w) - g(\hat{w}) = \int_{\hat{w}}^{w} g'(\zeta)\, d\zeta$).

Lemma 5.3.4. *Fixing some base point $\hat{w}_0$ in the half-plane $\mathbb{R}e(w) > R$, we define ϕ_k by*

$$\phi_k(w) = g^k(w) - g^k(\hat{w}_0),$$

for all $\mathbb{R}e(w) > R$ and $k \geq 1$. The following statements hold:

(i) *The sequence (ϕ_k) converges uniformly on compact sets of the half-plane $\mathbb{R}e(w) > R$ to an analytic and univalent function ϕ satisfying the equality $\phi \circ g = \phi + 1$.*

(ii) *One has*

$$\lim_{k\to\infty} \frac{g^k(w) - g^k(\hat{w}_0)}{w - \hat{w}_0} = \frac{\phi(w)}{w - \hat{w}_0}$$

uniformly throughout the half-plane $\mathbb{R}e(w) > R$.

Proof. (i) In particular, it follows from Lemma 5.3.2 that

$$\left|\frac{g^k(w) - g^k(\hat{w})}{g^{k-1}(w) - g^{k-1}(\hat{w})}\right| \leq 1 + \frac{C'}{k^{1+1/n}}, \tag{5.26}$$

for all w and $\hat{w}$ in the half-plane H_R. To prove the convergence of (ϕ_k), we consider the series

$$w - \hat{w}_0 + \sum_{k\geq 1}\left[(g^k(w) - g^k(\hat{w}_0)) - (g^{k-1}(w) - g^{k-1}(\hat{w}_0))\right].$$

As the infinite product $P = \prod_{k\geq 1}(1 + \frac{C'}{k^{1+1/n}})$ is convergent, it follows from (5.26) that

$$|g^{k-1}(w) - g^{k-1}(\hat{w}_0)| \leq P|w - \hat{w}_0|, \tag{5.27}$$

for every k. Setting $\hat{w} = \hat{w}_0$ in (5.23) and multiplying the resulting inequality by (5.27), we obtain

$$|(g^k(w) - g^k(\hat{w}_0)) - (g^{k-1}(w) - g^{k-1}(\hat{w}_0))| \leq PC'|w - \hat{w}_0|/k^{1+1/n}. \tag{5.28}$$

This proves that (ϕ_k) converges uniformly on compact sets of H_R to an analytic function ϕ. Since all ϕ_k's are univalent, so is ϕ. Finally, let us show that $\phi \circ g = \phi + 1$. For $\mathbb{R}e(w) > R$, we have

$$\begin{aligned}\phi(g(w)) &= \lim_{k\to\infty}(g^{k+1}(w) - g^k(\hat{w}_0))\\ &= \lim_{k\to\infty}(g^{k+1}(w) - g^{k+1}(\hat{w}_0)) + \lim_{k\to\infty}(g^{k+1}(\hat{w}_0) - g^k(\hat{w}_0)) = \phi(w) + 1.\end{aligned}$$

The equality $\lim_{k\to\infty}(g^{k+1}(\hat{w}_0) - g^k(\hat{w}_0)) = 1$ is evident in view of (5.20) and (5.22).

(ii) It follows from (5.28) that

$$|F_k(w) - F_{k-1}(w)| \leq PC'|w - \hat{w}_0|/k^{1+1/n}, \tag{5.29}$$

where $F_k(w) = g^k(w) - g^k(\hat{w}_0) - \phi(w)$. For $M > N$, we have

$$F_M(w) - F_N(w) = \sum_{h=N+1}^{M} [F_h(w) - F_{h-1}(w)].$$

By (5.29), $(F_k(w)/(w - \hat{w}_0))_k$ converges uniformly in $\mathbb{R}e(w) > R$. □

Lemma 5.3.5. *For w in the half-plane $H_R = \{w : \mathbb{R}e(w) > R\}$ we have* $\lim_{w\to\infty} \phi(w)/(w - \hat{w}_0) = 1$.

Proof. In view of the inequality

$$\left|\frac{\phi(w)}{w - \hat{w}_0} - 1\right| \leq \left|\frac{\phi(w)}{w - \hat{w}_0} - \frac{g^k(w) - g^k(\hat{w}_0)}{w - \hat{w}_0}\right| + \left|\frac{g^k(w) - g^k(\hat{w}_0)}{w - \hat{w}_0} - 1\right|$$

and statement (ii) in Lemma 5.3.4, it suffices to show that $\lim_{w\to\infty} g^k(w) - w = k$ for every $k \in \mathbb{N}$. For this, write (5.19) in the form

$$g(w) = 1 + w + r(w), \quad \text{where } \lim_{w\to\infty} r(w) = 0.$$

For a fixed k, it follows from the above relation that

$$g^k(w) = w + k + \sum_{0}^{k-1} r(g^j(w))$$

which proves that $\lim_{w\to\infty} g^k(w) - w = k$ for every k. Notice that, in particular, the asymptotic equality $\phi(w) \sim w$ holds. □

The set $\phi(H_R)$ contains a right half-plane $H = \{z : \mathbb{R}e(z) > c\}$ where c is a positve constant. A proof of this result that uses Rouche's Theorem and the asymptotic equality $\phi(w) \sim w$ can be consulted in ([55], Section 15.1).

Theorem 5.3.6. *Let f be an entire function and z^* a fixed point of f with multiplier $\lambda = 1$. If P is an attracting petal, there is a conformal map α defined in P satisfying the Abel equation*

$$\alpha(f(z)) = \alpha(z) + 1 \quad \textit{for all } z \in P. \tag{5.30}$$

With a suitable choice of P, $\alpha(P)$ will contain some right half-plane $\{w : \mathbb{R}e(w) > c\}$.

Proof. In Lemma 5.3.4, we have proved that the function defined by $\phi(w) = \lim_{k\to\infty}(g^k(w) - g^k(\hat{w}_0))$ for all $w \in H_R$ is univalent and satisfies the Abel equation

$\phi(g(w)) = \phi(w)+1$. Therefore, setting $\alpha(z) = \phi(Tz)$, the univalent function α is defined in the attracting petal $P_R = T^{-1}\{w : \mathbb{R}e(w) > R\}$ and verifies the equation $\alpha(f(z)) = \alpha(z) + 1$ for all $z \in P_R$.

Now, given an arbitrary attracting petal P, we will prove that α can be extended to the entire P. Indeed, if $z \in P$, there is a $k \in \mathbb{N}$ such that $f^k(z) \in P_R$. Choose a disc D centered at $f^k(z)$ satisfying $D \subset P_R$. By continuity, there exists a disc E with center z such that $f^k(E) \subset D$. For all $z \in E$, we put $\alpha(z) = \alpha(f^k(z)) - k$ (notice that $\alpha(f^k(z)) - k = \alpha(f^h(z)) - h$ for all $h > k$). Obviously, α is analytic and, using that f is univalent in $P \subset W \cap W'$, it is easy to deduce that it is still univalent.

To show that $\alpha(P_R)$ contains a right half-plane, notice that $\alpha(P_R) = \phi \circ T(P_R) = \phi \circ T \circ T^{-1}(H_R) = \phi(H_R)$. □

The function α is often referred to as a *Fatou function* for the petal P and $\alpha(z)$ as the Fatou coordinate in P. A Fatou function is uniquely determined up to an additive constant. We omit its proof, the interested reader may consult [61].

Remarks 5.3.7. (1) Function f is conjugate to a translation $T : \zeta \to \zeta + 1$ in some subregion of an attracting petal P: Let α be a Fatou function satisfying $\alpha(f(z)) = \alpha(z) + 1$. We know that $\alpha(P)$ contains a right half-plane $H = \{w : \mathbb{R}e(w) > c\}$. If we put $P_0 = \alpha^{-1}(H)$, notice that $f(P_0) \subset P_0$ and the following commutative diagram holds:

$$\begin{array}{ccc} P_0 & \xrightarrow{f} & P_0 \\ \alpha\downarrow & & \downarrow\alpha \\ H & \xrightarrow{T} & H \end{array}$$

(2) As in the attracting case, the Fatou function may be extended to a function defined and analytic in the basin of attraction of P, still satisfying the Abel equation.

(3) An important consequence of the parabolic linearization that we will need later is the following fact. Under the above conditions, the iterates of every $z \in P$ are different. Indeed, suppose $f^n(z) = f^m(z)$ for some $n > m$. Applying (5.30), we obtain $\phi(f^n(z)) = \phi(f^m(z)) + n - m$, which is a contradiction.

(4) If P is an attracting petal for f, we know that $f(P) \subset P$. Let us see that the inclusion is strict. In fact, the equality $f(P) = P$ implies that $\alpha(P) = \alpha(P) + 1$, with α being a Fatou function for the petal P. Since $\alpha(P)$ contains a right half-plane, it follows that $\alpha(P) = \mathbb{C}$. As α is univalent, this shows that P is conformally equivalent to $\mathbb{C}$. By Riemann's theorem, P is conformally equivalent to the unit disc $\mathbb{D}$, thus we have reached a contradiction.

5.4 Irrationally indifferent points

Now we consider the problem of the existence of linearization in a neighborhood of an indifferent fixed point z^*. The existence of linearization depends on f and, in particular, on the multiplier $\lambda = f'(z^*)$. In the case of a rationally indifferent point, the existence of attracting petals is incompatible with the linearization. Then, we only need to consider irrationally indifferent points. Recall that, in this case, the multiplier can be expressed in the form $\lambda = \exp(2\pi\theta i)$, with $\theta \notin \mathbb{Q}$. If there exists a linearization, the fixed point is called a Siegel point. Otherwise, it is called a Cremer point.

If $f(z)$ can be linearizated, that is, if f is conjugate to $g(w) = \lambda w$ in a neighborhood U of z^*, then U cannot contain periodic points of $f(z)$ different from the fixed point z^*. In the next chapter, we will see that the Julia set of f, $\mathbb{J}(f)$, is the closure of the set formed for all periodic repelling points (this is the so-called first fundamental theorem of dynamics). Therefore, the neighborhood U must be contained in the complement of $\mathbb{J}(f)$ (the Fatou set of f).

Let us see a brief history of this problem. In 1912, E. Kasner established the conjecture that linearization is always possible. But five years later, G. A. Pfeiffer proved that Kasner was wrong because he proved that for certain functions the linearization is not possible. In 1919, Julia believed to have proved that the linearization never is possible, but his proof is incorrect. In 1927, in a beautiful work, H. Cremer proved that for a generic choice of λ in the unit disc, there is no linearization. Concretely, he proved that, if f is rational of degree $d \geq 2$, then in a neighborhood of a neutral fixed point there are infinitely many periodic points of f whenever the multiplier $\lambda = e^{2\pi\theta i}$ has the property

$$\liminf_{q\to\infty} |\lambda^q - 1|^{1/d^q} = 0.$$

In this section, we will follow mainly [20].

Throughout the section, θ is an irrational number belonging to $[0,1]$, $\lambda = e^{2\pi\theta i}$, and f is an analytic function at the origin such that 0 is a fixed point with multiplier λ. If we put $h = \phi^{-1}$ in Schröder's equation

$$\phi(f(z)) = \lambda\phi(z),$$

the equation adopts the form

$$f(h(\zeta)) = h(\lambda\zeta) \quad \text{normalized by } h'(0) = 1. \tag{5.31}$$

The proof of the following theorem will be left to the reader as Exercise 2.

Theorem 5.4.1. *A solution h of the above equation in $\{\zeta : |\zeta| < r\}$ is univalent.*

Theorem 5.4.2. *The Schröder equation* (5.31) *has a solution if and only if (f^n) is uniformly bounded in a neighborhood of the origin.*

Proof. If h is a solution of (5.31) in $\mathbb{D}_r$, then

$$f^n(z) = h(\lambda^n h^{-1}(z)). \tag{5.32}$$

Choose an $s \in (0,r)$ and set $U = h(\mathbb{D}_s)$. If $K = \{\lambda^n h^{-1}(z) : n \in \mathbb{N}, z \in U\}$, it is obvious that $K \subset \mathbb{D}_s$. As h is continuous in $\overline{\mathbb{D}_s}$, there exists an $M > 0$ such that $|h(\lambda^n h^{-1}(z))| \leq M$ for all $z \in U$, and (5.32) yields the necessity.

In the opposite direction, suppose $|f^n(z)| \leq M$ for all $z \in U$ and $n \in \mathbb{N}$, with U being a neighborhood of the origin. Define ϕ_n by

$$\phi_n(z) = \frac{1}{n} \sum_{j=0}^{n-1} \lambda^{-j} f^j(z).$$

Notice that $|\phi_n(z)| \leq M$ for z in U, thus (ϕ_n) admits a subsequence that is uniformly convergent to a function ϕ analytic in U. On the other hand, let us see the form of $\phi_n \circ f$,

$$\phi_n(f(z)) = \frac{1}{n} \sum_{j=0}^{n-1} \lambda^{-j} f^{j+1}(z) = \lambda \phi_n(z) + \frac{1}{n}(\lambda^{-n} f^n(z) - \lambda z).$$

Since the last term on the right-hand side is uniformly convergent to 0 on U, the limit function ϕ verifies

$$\phi(f(z)) = \lambda \phi(z), \quad \text{for all } z \in U.$$

Finally, as $f'(0) = \lambda$, it follows from the definition of ϕ_n that $\phi_n'(0) = 1$ for all $n \in \mathbb{N}$. Therefore, $\phi'(0) = 1$ and $h = \phi^{-1}$ is the required solution of (5.31). □

Remarks 5.4.3. (1) As a consequence of the above theorem, we will prove that if f is topologically conjugate to λz, then it is conformally conjugate. Let h be a homeomorphism satisfying $h(0) = 0$ and $h^{-1} \circ f \circ h(\zeta) = \lambda \zeta$ for $\zeta \in \overline{\mathbb{D}}_\delta$. We claim that f is invariant on $U = h(\overline{\mathbb{D}}_\delta)$. In fact, if $z \in U$, then there exists $\zeta \in \overline{\mathbb{D}}_\delta$ such that

$$f(z) = f(h(\zeta)) = h(\lambda \zeta) \in U.$$

Then $f^n(U) \subset U$ for all $n \in \mathbb{N}$ and this implies that (f^n) is uniformly bounded in U. An application of Theorem 5.4.2 concludes the proof.

(2) It is well known that a locally uniformly bounded family of analytic functions is normal. Therefore, by Theorem 5.4.2, if z^* is a Siegel fixed point for f, then (f^n) is normal in a neighborhood of z^*.

Theorem 5.4.4. *There exists a* $\lambda = e^{2\pi\theta i}$ *such that Schröder's equation* (5.31) *has no solution for every polynomial* f.

Proof. Let $f(z) = z^d + \cdots + \lambda z$ be a polynomial of degree d such that the origin is an irrational indifferent fixed point with multiplier $\lambda = e^{2\pi\theta i}$ and suppose that equation (5.31) has a solution h in $\overline{\mathbb{D}}_\delta$. The equation

$$f^n(z) - z = z^{d^n} + \cdots + (\lambda^n - 1)z = 0$$

has d^n solutions and these are precisely the fixed points of f^n. The origin is a simple zero because $\lambda^n \neq 1$ and we denote the others by $z_1, z_2, \ldots, z_{d^n-1}$. We set $U = h(\overline{\mathbb{D}}_\delta)$ and will prove that $z_j \notin U$ for all $j \leq d^n - 1$. If $z_j \in U$, then $z_j = f^n(z_j) = h(\lambda^n h^{-1}(z_j))$, implying that $h^{-1}(z_j) = \lambda^n h^{-1}(z_j)$. So, $h^{-1}(z_j) = 0$, which is not possible because h is a bijective map from $\overline{\mathbb{D}}_\delta$ onto U. If we choose an $s > 0$ such that $\mathbb{D}_s \subset U$, then $|z_j| \geq s$ for $j \leq d^n - 1$ and

$$s^{d^n-1} \leq \prod_{j=1}^{d^n-1} |z_j| = |\lambda^n - 1|. \tag{5.33}$$

For the latter equality, note that there is only a nonzero product of $d^n - 1$ zeros of the polynomial $f^n(z) - z$ and its value is, except for the sign, the coefficient of z. Now the key is to determine λ such that (5.33) does not hold. Let (q_n) be a strictly increasing sequence of positive integers and put $\theta = \sum_n 2^{-q_n}$ and $\lambda = e^{2\pi\theta i}$. We will prove that

$$\left|1 - \lambda^{2^{q_k}}\right| \leq 4\pi\, 2^{q_k - q_{k+1}}. \tag{5.34}$$

First, elementary calculus yields

$$\left|1 - \lambda^{2^{q_k}}\right| = 2\left|\sin(\pi\theta 2^{q_k})\right| = 2\left|\sin\left(\pi\left(\sum_{i\geq 1} 2^{q_k - q_{k+i}}\right)\right)\right|$$

(notice that $\sin(\pi(\sum_{i\leq k} 2^{q_k - q_i})) = 0$). Let us obtain an upper bound of the sum of the latter series

$$\sum_{i\geq 1} 2^{q_k - q_{k+1}} = 2^{q_k} \sum_{i\geq 1} 2^{q_{k+1}} \leq \sum_{j\geq q_{k+1}} \frac{2^{q_k}}{2^j} = 2(2^{q_k - q_{k+1}}).$$

The relation (5.34) follows because $|\sin x| \leq |x|$ for all real x. Applying (5.33) and (5.34), for each $n = 2^{q_k}$ one gets

$$s^{d^{2^{q_k}}-1} \leq 4\pi\, 2^{q_k - q_{k+1}}. \tag{5.35}$$

Finally, multiplying by s both sides and taking the logarithm, we deduce

$$d^{2^{q_k}} \log_2(s) \leq \log_2(4\pi s) + q_k - q_{k+1},$$

which implies

$$q_{k+1} \leq \log_2(4\pi s) + q_k + d^{2^{q_k}} \log_2(1/s).$$

Therefore, it suffices to construct inductively a sequence (q_k) satisfying

$$q_{k+1} > q_k + q_k^{2^{q_k}}$$

to reach a contradiction. □

We omit the easy proof of the following theorem (Exercise 5).

Theorem 5.4.5 (Small cycles, Cremer). *Let $f(z)$ be a polynomial function of degree $d \geq 2$ and $\lambda \neq 1$ a complex number on the unit circle such that*

$$\liminf_{q\to\infty} |\lambda^q - 1|^{1/d^q} = 0.$$

If the origin is a fixed point for f with multiplier λ, then in every neighborhood of zero there exist periodic points $z \neq 0$. Therefore, f is not linearizable at the origin.

In view of the above theorem, the natural question is whether there are irrational numbers θ satisfying

$$\liminf_{q\to\infty} |e^{2\pi q\theta i} - 1|^{1/d^q} = 0.$$

To answer this question, we need to consider the notions of continued fractions and diophantine numbers.

Definition 5.4.6. Given $\kappa > 0$, a real number θ is called diophantine of order κ if it is not approximable by rational numbers in the sense that there exists an $\epsilon > 0$ such that

$$\left|\theta - \frac{p}{q}\right| > \frac{\epsilon}{q^\kappa} \quad \text{for all } p, q \in \mathbb{Z},\ q > 0. \tag{5.36}$$

We denote by $D(\kappa)$ the set formed by all diophantine numbers of order κ.

We are going to prove that (5.36) is equivalent to the following condition:

$$|\lambda^q - 1| > \frac{\epsilon'}{q^{\kappa-1}}, \tag{5.37}$$

for some $\epsilon' > 0$ and $q \in \mathbb{N}$. First, for $\lambda = e^{2\pi\theta i}$ we have

$$|\lambda^q - 1|^2 = 2(1 - \cos(2\pi\theta q)) = 4\sin^2(\pi\theta q),$$

so

$$|\lambda^q - 1| = 2|\sin(\pi\theta q)| = 2|\sin(\pi(q\theta - p))|,$$

for all $p \in \mathbb{Z}$ and $q \in \mathbb{N}$.

For a fixed $q \in \mathbb{N}$, if p is the closest integer to $q\theta$, then $|\theta q - p| \leq 1/2$. Since $|\sin y| \geq (2/\pi)|y|$ for $y \in [-\pi/2, \pi/2]$, it follows that

$$|\lambda^q - 1| \geq 4|q\theta - p|.$$

On the other hand, $|\sin y| \leq |y|$ for $y \in [-\pi/2, \pi/2]$, thus

$$|\lambda^q - 1| \leq 2\pi|\theta q - p|.$$

Therefore, we have

$$4|q\theta - p| \leq |\lambda^q - 1| \leq 2\pi|\theta q - p|, \tag{5.38}$$

for each $q \in \mathbb{N}$, where p is the integer closest to θq. From (5.38), it is easy to deduce the equivalence of (5.36) and (5.37).

The next theorem is an interesting classical result about diophantine numbers (Exercise 3).

Theorem 5.4.7 (Liouville). *If a real number θ satisfies a polynomial equation $f(x) = 0$ of degree d with integer coefficients, then θ is diophantine of order d.*

The following result states that almost all real numbers are diophantine.

Theorem 5.4.8. *If $\kappa > 2$, then the set of all diophantine numbers in the interval $[0,1]$ has measure 1.*

Proof. Let $U(\epsilon)$ be the set of all $\theta \in [0,1]$ such that there exists a p/q satisfying $|\theta - p/q| < \epsilon/q^\kappa$. The measure of this set is less than or equal to

$$\sum_{q=1}^{\infty} q\frac{2\epsilon}{q^\kappa},$$

because for each q there are q different forms for p/q. Since $\kappa > 2$, the above series is convergent and its sum converges to 0 as $\epsilon \to 0$. Thus the set $\bigcap_{\epsilon>0} U(\epsilon)$ has measure 0 and, therefore, its complement in $[0,1]$ has measure 1. Notice that this set is precisely $D(\kappa)$. □

Theorem 5.4.9 (Siegel). *If θ is diophantine, then every analytic function having a fixed point with multiplier $\lambda = e^{2\pi\theta i}$ is locally linearizable.*

The proof has rather technical complications and we omit it. For a proof that uses the KAM theory (by Kolmogorov, Arnold, and Moser), the interested reader may consult [20].

5.5 Continued fractions

In this final section, we will study the classical simple continued fraction algorithm. As we will see, this subject is closely related with the notion of the best rational approximation to an irrational number.

Given a real number $0 < \theta < 1$, we define inductively two sequences (θ_n) and (a_n) such that

$$\frac{1}{\theta_{n-1}} = \theta_n + a_n, \tag{5.39}$$

where $a_n = \lfloor 1/\theta_{n-1} \rfloor$ and $\theta_0 = \theta$. Then, the a_n's are positive integers and $0 \le \theta_n < 1$. This process may be summarized in the following way (infinite continued fraction):

$$\theta = \frac{1}{a_1 + \theta_1} = \frac{1}{a_1 + \frac{1}{a_2+\theta_2}} = \cdots = \cfrac{1}{a_1 + \cfrac{1}{a_2 + \cfrac{1}{a_3 + \cfrac{1}{a_4 + \ddots}}}}.$$

In the above equalities, the final term is called the continued fraction expansion of θ. We denote by $[a_1, a_2, \ldots, a_n]$ the finite continued fraction given by

$$\cfrac{1}{a_1 + \cfrac{1}{a_2 + \cfrac{1}{\ddots \; \cfrac{1}{a_{n-2} + \cfrac{1}{a_{n-1} + \cfrac{1}{a_n}}}}}}.$$

Usually, $[a_1, a_2, \ldots, a_n]$ is called the nth convergent.

Under the above conditions, we define inductively two sequences of nonnegative integers (p_n) and (q_n) as follows:

$$\begin{cases} p_0 = 1, \; p_1 = 0, \; p_{n+1} = a_n p_n + p_{n-1}, \\ q_0 = 0, \; q_1 = 1, \; q_{n+1} = a_n q_n + q_{n-1}. \end{cases} \tag{5.40}$$

It is easy to prove the inequalities

$$p_{n+1} \ge 2p_{n-1} \quad \text{and} \quad q_{n+1} \ge 2q_{n-1}.$$

Therefore, the sequences (p_n) and (q_n) tend to infinity. The equations in (5.40) can be written using matrix notation as

$$\begin{bmatrix} p_n & q_n \\ p_{n+1} & q_{n+1} \end{bmatrix} = \begin{bmatrix} 0 & 1 \\ 1 & a_n \end{bmatrix} \begin{bmatrix} p_{n-1} & q_{n-1} \\ p_n & q_n \end{bmatrix}.$$

Applying repeatedly this relation, we obtain

$$\begin{bmatrix} p_n & q_n \\ p_{n+1} & q_{n+1} \end{bmatrix} = \begin{bmatrix} 0 & 1 \\ 1 & a_n \end{bmatrix} \begin{bmatrix} 0 & 1 \\ 1 & a_{n-1} \end{bmatrix} \cdots \begin{bmatrix} 0 & 1 \\ 1 & a_1 \end{bmatrix}.$$

Thus the determinant of the matrix on the left is given by

$$p_n q_{n+1} - q_n p_{n+1} = (-1)^n. \tag{5.41}$$

In particular, it follows that p_n and q_n are relatively prime, and each p_n/q_n is a fraction in lowest terms. Now setting $d_n = \theta_0\theta_1 \cdots \theta_{n-1}$, the equation (5.39) multiplied by d_n takes the form

$$d_{n-1} = a_n d_n + d_{n+1}, \tag{5.42}$$

with $d_0 = 1$ and $d_1 = \theta_0$. If we put $\varepsilon_n = (-1)^n d_n$, then (5.42) adopts the form

$$\varepsilon_{n+1} = a_n \varepsilon_n + \varepsilon_{n-1}. \tag{5.43}$$

By induction and using (5.40) and (5.43), we deduce the equality

$$\varepsilon_n = p_n - q_n\theta_0. \tag{5.44}$$

In particular, we have $d_n = |p_n - q_n\theta|$. Then we can write

$$\theta = \frac{p_n}{q_n} - \frac{\varepsilon_n}{q_n}, \quad \text{where } \varepsilon_n = (-1)^n\theta_0\theta_1 \cdots \theta_{n-1}. \tag{5.45}$$

Proposition 5.5.1. *Under the above conditions, we have*
(i) $0 = \frac{p_1}{q_1} < \frac{p_3}{q_3} < \cdots < \theta_0 < \cdots < \frac{p_4}{q_4} < \frac{p_2}{q_2} < 1.$
(ii) $p_n/q_n \to \theta$ *as* $n \to \infty$.

Proof. (i) In view of (5.45), it is evident.

(ii) Again by (5.45), in the approximation $\theta \approx p_n/q_n$, the error is $d_n/q_n = \varepsilon_n/q_n$. Using (5.41) and (5.45), we can estimate its size as follows:

$$\frac{d_{n+1}}{q_{n+1}} + \frac{d_n}{q_n} = \left|\frac{p_{n+1}}{q_{n+1}} - \theta\right| + \left|\theta - \frac{p_n}{q_n}\right| = \left|\frac{p_{n+1}}{q_{n+1}} - \frac{p_n}{q_n}\right| = \frac{1}{q_n q_{n+1}}.$$

In particular, we have obtained

$$\frac{d_n}{q_n} < \frac{d_{n+1}}{q_{n+1}} + \frac{d_n}{q_n} = \frac{1}{q_n q_{n+1}},$$

or, equivalently,

$$d_n < \frac{1}{q_{n+1}}.$$

Then we deduce that $d_n \to 0$ and p_n/q_n converges to θ. ☐

Next, we establish the relationship between diophantine numbers of order κ and the boundedness of the sequence (q_{n+1}/q_n^κ). The proof will be left to the reader (Exercise 7).

Corollary 5.5.2. *An irrational number θ is diophantine of order κ if and only if the sequence $(q_{n+1}/q_n^{\kappa-1})$ is bounded.*

Finally, we consider briefly the problem of the best approximation of irrational numbers by rational numbers.

Definition 5.5.3. Let θ be an irrational number and $a/b \in \mathbb{Q}$. If

$$\left|\theta - \frac{a}{b}\right| < \left|\theta - \frac{c}{d}\right|$$

for all $0 < d \leq b$, we will say that a/b is a best approximation of type 1 to the irrational number θ.

In the same terms, if $|b\theta - a| < |d\theta - c|$, we will say that the rational number a/b is a best approximation of type 2 to θ.

Proposition 5.5.4. *Every best approximation of type 2 of a real number θ is a best approximation of type 1 of θ.*

Proof. Let a/b be a best approximation of type 2 to θ. If c/d is a rational number with $0 < d \leq b$, then we have

$$\left|\theta - \frac{a}{b}\right| = \frac{|b\theta - a|}{b} < \frac{|d\theta - c|}{b} \leq \frac{|d\theta - c|}{d} = \left|\theta - \frac{c}{d}\right|.$$

□

The proof of the next theorem may be consulted in [77, p. 20].

Theorem 5.5.5. *Every best approximation of type 2 to an irrational number θ is a convergent of its continued fraction expansion.*

5.6 Exercises

1. The origin is a repelling fixed point of $f(z) = 2z + 2z^2$. Show that f is conjugate to $L(z) = 2z$ trhough the function $h(z) = \log(1 + 2z)$ in a neighborhood of zero, where log is the principal branch of the logarithm.
2. Consider the equation

$$f(h(\zeta)) = h(\lambda\zeta) \quad \text{normalized by } h'(0) = 1.$$

   Prove that if h is a solution of the above equation in $\{\zeta : |\zeta| < r\}$, then h is univalent.
3. Prove Liouville's theorem: If the real number θ satisfies a polynomial equation $f(x) = 0$ of degree d with integer coefficients, then θ is diophantine of order d.

4. Show that, for any irrational number x, there are infinitely many fractions p/q with

$$\left|x - \frac{p}{q}\right| < \frac{1}{q^2}.$$

5. Prove the small cycles theorem (Cremer).
6. With the notations of the last section, prove the following statements:
   (i) If $\theta = p/q$ is rational, then its continued fraction expansion is finite.
   (ii) $[a_1, a_2, \dots, a_n] = \frac{p_n}{q_n}$ for all $n \geq 1$.
7. Prove Corollary 5.5.2.
8. Show that $\frac{13}{4}$ is a best approximation of type 1 to π, but it is not a best approximation of type 2.
9. Consider the function $f(z) = \lambda z + z^2$, with λ being a complex constant. Determine a formal power series $z + \sum_{n\geq 2} a_n z^n$ that satisfies Schröder's equation $\phi(f(z)) = \lambda\phi(z)$. Describe the set of all values of λ for which there is no such a series.
10. Let f be an entire function with a Siegel point z_0. Prove that there is a connected neighborhood U of z_0 with the following property: "For every $z \in U$ $(z \neq z_0)$, there exists a Jordan curve γ in U such that the orbit $O^+(z)$ is a dense subset of γ".

Bibliography remarks

In Section 5.1, we follow [20, Section II.2].

Section 5.2 is based on [13, Section 6.10] and [20, Section II.4].

Section 5.3 is inspired by [61, Section 10], [20, Section III.5], and [75, Section 3.5], in this order.

Section 5.4 is based on [20, Section III.6].

In Section 5.5, we follow [61, Appendix C].

6 The Fatou and Julia sets

6.1 Definitions and examples

Definition 6.1.1. For any entire function $f \in \mathcal{E}$, the Fatou set (or stable set) of f is formed by all z_0 such that (f^n) is normal in a neighborhood of z_0 and it is denoted by $\mathbb{F}(f)$. Its complement in $\mathbb{C}$, denoted by $\mathbb{J}(f)$, is called the Julia set of f. It is evident that $\mathbb{F}(f)$ is open and, therefore, $\mathbb{J}(f)$ is a closed set.

In the framework of rational functions, these definitions of $\mathbb{F}(f)$ and $\mathbb{J}(f)$ are due to Fatou. Julia defined $\mathbb{J}(f)$ as the closure of the set of all repelling periodic points of f. A central result in both works is the equivalence of the two definitions. Anyway, the usual practice is to adopt Fatou's definition. The following result is useful to obtain computer generated graphs of Julia sets.

Theorem 6.1.2. *If $\mathbb{B}(f)$ denotes the set of all points $z_0 \in \mathbb{C}$ with a bounded orbit, then*

$$\partial(\mathbb{B}(f)) \subset \mathbb{J}(f).$$

Proof. Take $z_0 \in \partial(\mathbb{B}(f))$ such that $z_0 \notin \mathbb{J}(f)$. Then (f^n) is normal in a compact neighborhood U of z_0. Since z_0 is a boundary point of $\mathbb{B}(f)$, there are points $z_1, z_2 \in U$ such that the orbit of z_1 is bounded while the orbit of z_2 is not. Thus $(f^n(z_2))$ admits a divergent subsequence $(f^{n_k}(z_2))_k$. By the normality of (f^n) in U, there is a subsequence of $(f^{n_k}(z))_k$ uniformly convergent (or divergent) on U. Thus we have reached a contradiction because $(f^n(z_1))$ is bounded and $(f^{n_k}(z_2))_k \to \infty$. □

The set $\mathbb{B}(f)$ is called the *filled Julia set* of f. As we will see later, the following equality:

$$\mathbb{J}(f) = \partial(\mathbb{B}(f))$$

holds for polynomial functions.

Remark 6.1.3. If f is a polynomial function of degree $n \geq 2$, in Section 6.5, we will prove that there exists an $R > 0$ such that the orbit of z tends to ∞ whenever $|z| > R$. Therefore, ∞ is an attracting fixed point for f since $f(\infty) = \infty$. That is the reason why the dynamical study of these functions must be carried out in the extended plane.

Examples 6.1.4. 1) Let $f(z) = z^2$. In Example 1.2.1(2), we have seen that
- $\lim_{n\to\infty} f^n(z) = 0$, if $|z| < 1$.
- $\lim_{n\to\infty} f^n(z) = \infty$, if $|z| > 1$.

Then the boundary of $\mathbb{B}(f)$ is the unit circle $|z| = 1$. Thus Theorem 6.1.2 tells us that this circle is contained in $\mathbb{J}(f)$. On the other hand, the origin is an attracting fixed point of f and $|z| < 1$ is its attractive basin. As we will see in Theorem 6.2.2, this basin is contained in the Fatou set of f. Finally, note that $|z| > 1$ is the attractive basin of ∞ and, again by Theorem 6.2.2, we conclude that it is contained in $\mathbb{F}(f)$. Thus $\mathbb{J}(f)$ is the unit circle.

https://doi.org/10.1515/9783111689685-006

2) Let $g(z) = z^2 - 2$. We will prove that $\mathbb{J}(g) = [-2, 2]$. This $g(z)$ is conjugate to $f(z) = z^2$ through the map $h(z) = z + 1/z$, which is a conformal isomorphism from $A = \{z : |z| > 1\}$ onto $B = \mathbb{C} \setminus [-2, 2]$ (see Exercise 1.2). As h is analytic in A and $h'(z) \neq 0$, the inverse function theorem tells us that h^{-1} is analytic.

Now we will see that f and g are conjugate:

$$g(h(z)) = h(z)^2 - 2 = \left(z + \frac{1}{z}\right)^2 - 2 = z^2 + \frac{1}{z^2} \quad \text{and} \quad h(f(z)) = f(z) + \frac{1}{f(z)} = z^2 + \frac{1}{z^2}.$$

Since $\lim_{n\to\infty} f^n(z) = \infty$ for $|z| > 1$, it follows that $\lim_{n\to\infty} g^n(z) = \infty$ for $z \in B$. To conclude that $\mathbb{J}(g) = [-2, 2]$, notice that $h(z)$ maps the unit circle onto the interval $[-2, 2]$.

The quadratic family $Q_c(z) = z^2 + c$ will be considered in Section 6.6. We have just studied the cases $c = 0, -2$ in the above example. In both of them, the Julia set is very simple; however, in general, the Julia set of Q_c presents an intricate fractal geometry for complex values of c.

Theorem 6.1.5. *The sets $\mathbb{J}(f)$ and $\mathbb{F}(f)$ are completely invariant.*

Proof. As $\mathbb{J}(f)$ and $\mathbb{F}(f)$ are complementary sets, we only need to prove that $\mathbb{J}(f)$ is completely invariant. Let $z_0 \in \mathbb{J}(f)$ and, by contradiction, suppose that $f(z_0) \notin \mathbb{J}(f)$. This means that there exists a neighborhood U of $f(z_0)$ such that $(f^n)_n$ is normal in U. We put $V = f^{-1}(U)$ and will prove that $(f^n)_n$ is normal in the neighborhood V of z_0, which is a contradiction. In fact, given a sequence $(f^{n_k})_k$, consider the sequence $(f^{n_k-1})_k$. By normality, there exists a subsequence that is uniformly convergent on compact subsets of U to an analytic function g (or to ∞) (for the sake of simplicity, we go on denoting the subsequence by $(f^{n_k-1})_k$). Then, in the first case, for all $z \in V$, we have

$$\lim_{k\to\infty} f^{n_k-1}(f(z)) = g(f(z)) \quad \text{uniformly on compact subsets of } V.$$

Thus $(f^{n_k})_k$ is uniformly convergent on compact subsets of V to $g \circ f$ (the argument is similar when the convergence is to ∞). This proves that $f(z_0) \in \mathbb{J}(f)$.

Using the fact that f is open, the proof of the implication $f(z_0) \in \mathbb{J}(f) \Rightarrow z_0 \in \mathbb{J}(f)$ is similar. □

Sometimes, to simplify, we will denote the Julia and Fatou sets of f by $\mathbb{J}$ and $\mathbb{F}$, respectively.

It is well known, see (Julia, 1918 [51]), that every rational function f with degree $d \geq 2$ satisfies the following properties:

$$\mathbb{F} = f^{-1}(\mathbb{F}) = f(\mathbb{F}) \quad \text{and} \quad J = f^{-1}(\mathbb{J}) = f(\mathbb{J}),$$

(see, for example, [51]). Nevertheless, the next proposition shows that, for transcendental entire functions, the situation is different due to the existence of Picard exceptional values.

If f is a transcendental entire function, we will say that $a \in \mathbb{C}$ is a *Picard exceptional value* of f if the equation $f(z) - a = 0$ has no solution. We will denote the set of all finite Picard exceptional values of f by $PV(f)$. By the Little Picard theorem, we know that it contains at most one element.

Proposition 6.1.6. *Let f be a transcendental entire function with a Picard exceptional value a. The following equalities hold:*

$$\mathbb{F} = f^{-1}(\mathbb{F}) = f(\mathbb{F}) \cup (\{a\} \cap \mathbb{F}) \quad \text{and} \quad \mathbb{J} = f^{-1}(\mathbb{J}) = f(\mathbb{J}) \cup (\{a\} \cap \mathbb{J}).$$

If f has no Picard exceptional value, then

$$\mathbb{F} = f^{-1}(\mathbb{F}) = f(\mathbb{F}) \quad \text{and} \quad \mathbb{J} = f^{-1}(\mathbb{J}) = f(\mathbb{J}).$$

Proof. We have $A \subset f^{-1}(f(A))$ for all sets $A \subset \mathbb{C}$. This, together with Theorem 6.1.5, yields the inclusion

$$\mathbb{F} \subset f^{-1}(f(\mathbb{F})) \subset f^{-1}(\mathbb{F})$$

and, therefore, we have proved the equality $\mathbb{F} = f^{-1}(\mathbb{F})$. On the other hand, if $\mathbb{F}$ does not contain any Picard exceptional value of f, given $z_0 \in \mathbb{F}$, there exists a $z_1 \in \mathbb{C}$ such that $f(z_1) = z_0$. Again by Theorem 6.1.5, $z_1 \in \mathbb{F}$ and, in consequence, $z_0 = f(z_1) \in f(\mathbb{F})$. This proves that $\mathbb{F} \subset f(\mathbb{F})$, and concludes the proof in the case $PV(f) \cap \mathbb{F} = \emptyset$.

Now suppose that f has a Picard exceptional value $a \in \mathbb{F}$. We will show that $\{a\} = \mathbb{F}/f(\mathbb{F})$. This a belongs to $\mathbb{F}/f(\mathbb{F})$ because it is an exceptional value of f. Finally, if $z \in \mathbb{F}/f(\mathbb{F})$ with $z \neq a$, a fortiori there exists a w such that $z = f(w)$. By Theorem 6.1.5, it follows that $w \in \mathbb{F}$ and, therefore, $z = f(w) \in f(\mathbb{F})$, which is a contradiction. This implies $\{a\} = \mathbb{F}/f(\mathbb{F})$ and, therefore, $\mathbb{F} = f(\mathbb{F}) \cup \{a\}$. □

6.2 Properties of $\mathbb{J}(f)$ and $\mathbb{F}(f)$

Theorem 6.2.1. *Let f be a transcendental entire function. For every $k \in \mathbb{N}$,*

$$\mathbb{F}(f^k) = \mathbb{F}(f) \quad \text{and} \quad \mathbb{J}(f^k) = \mathbb{J}(f).$$

Proof. Obviously, it suffices to consider the case $k = 2$. The inclusion $\mathbb{F} \subset \mathbb{F}(f^2)$ is evident and, therefore, so we have $\mathbb{J}(f^2) \subset \mathbb{J}(f)$. To prove the converse inclusion $\mathbb{J}(f) \subset \mathbb{J}(f^2)$, take $z_0 \in \mathbb{J}(f)$ and suppose that $z_0 \notin \mathbb{J}(f^2)$. Then there exists a disk $\mathbb{D}_r(z_0)$ in which $(f^{2n})_n$ is normal. By Corollary 4.4.4 the function f^2 has infinitely many fixed points. As (f^n) is not normal in $\mathbb{D}_r(z_0)$, Montel's Theorem tells us that there exist a function f^p, a fixed point z^* of f^2, and a point η in $\mathbb{D}_r(z_0)$ such that $f^p(\eta) = z^*$ and $z^* = f^2(z^*)$. Now we will prove that the sequence $(f^{2n}(\eta))_n$ is bounded. For each integer n so that $2n \geq p$, the difference $2n - p$ is odd or even. Then there are two possibilities:

(i) There is an integer $m \geq 0$ such that $2n = 2m + p$. In this case, it holds

$$f^{2n}(\eta) = f^{2m} \circ f^{p}(\eta) = z^*.$$

(ii) There is an integer $m \geq 0$ satisfying $2n = 2m + p + 1$. Now we have

$$f^{2n}(\eta) = f \circ f^{2m} \circ f^{p}(\eta) = f(z^*).$$

This shows that $(f^{2n}(\eta))_n$ is bounded and since $(f^{2n})_n$ is normal in $\mathbb{D}_r(z_0)$ we deduce that it is locally uniformly bounded in $\mathbb{D}_r(z_0)$. Then take a constant $M > 0$ so that

$$|f^{2n}(z)| \leq M \quad \text{for all } n \in \mathbb{N} \text{ and } |z - z_0| < r/2.$$

Finally, notice that for each integer $n \geq 2$ there exist nonnegative integers m and s so that $n = 2m+s$ with $s = 0$ or $s = 1$. Then $f^n(z) = f^s \circ f^{2m}(z)$. From this it is easy to conclude that (f^n) is uniformly bounded in $|z - z_0| < r/2$ and then normal. □

Theorem 6.2.2. *If f is an entire function, the following statements hold:*
(1) *The attractive basin of all attracting cycle is contained in the Fatou set, while the boundary of that basin is contained in $\mathbb{J}(f)$.*
(2) *Every repelling cycle is contained in $\mathbb{J}(f)$.*
(3) *Every rationally indifferent fixed point belongs to $\mathbb{J}(f)$.*
(4) *Siegel points are in $\mathbb{F}(f)$ and Cremer points are in $\mathbb{J}(f)$.*

Proof. (1) As $\mathbb{J}(f) = \mathbb{J}(f^n)$, we only need to consider the case of a fixed point z_0.

If z_0 is an attracting fixed point of f, we know that $(f^n)_n$ converges uniformly to z_0 on compact subsets of the attractive basin $\Omega(f, z_0)$. Then (f^n) is normal in $\Omega(f, z_0)$.

Now let z^* a boundary point of $\Omega(f, z_0)$. By contradiction, suppose that $(f^n)_n$ is normal in a neighborhood U of z^* and choose z_1 in U such that $z_1 \notin \Omega(f, z_0)$. Then $(f^n(z_1))$ does not converge to z_0 and, therefore, there exist an $\epsilon > 0$ and a subsequence $(f^{n_k}(z_1))_k$ such that

$$|f^{n_k}(z_1) - z_0| > \epsilon \quad \text{for all } k \in \mathbb{N}. \tag{6.1}$$

Since $(f^n)_n$ is normal in U, a subsequence of $(f^{n_k})_k$ converges uniformly on compact subsets of U to an analytic function g. But $g(z) = z_0$ for all $z \in U \cap \Omega(f, z_0)$ and this set is infinite. So, the identity principle of analytic functions tells us that $g(z) = z_0$ for all $z \in U$. This contradicts (6.1).

(2) If z_0 is a repelling fixed point of f with multiplier $\lambda = f'(z_0)$, then

$$\frac{df^n}{dz}(z_0) = f'(z_0)^n = \lambda^n.$$

As $|\lambda| > 1$, one has $\lim_{n\to\infty}(f^n)'(z_0) = \infty$, however, $\lim_{n\to\infty} f^n(z_0) = z_0$. This means that $(f_n)_n$ cannot be normal at z_0.

(3) For the sake of simplicity, we will suppose that $z_0 = 0$ is a parabolic fixed point of f, then

$$f(z) = z + az^{k+1} + \cdots,$$

with $a \neq 0$. Elementary calculus shows that

$$f^p(z) = z + paz^{k+1} + \cdots.$$

Thus

$$\frac{d^{k+1}f^p}{dz^{k+1}}(0) = ap(k+1)!,$$

which implies that

$$\lim_{p\to\infty} \frac{d^{k+1}f^p}{dz^{k+1}}(0) = \infty.$$

Nevertheless, $f^p(0) = 0$ for all p. This shows that $(f_n)_n$ is not normal in a neighborhood of $z_0 = 0$.

Finally, if z_0 is a rationally indifferent fixed point, then $f'(z_0) = \lambda$, where λ is an mth root of unity. In this case, $(f^m)'(z_0) = \lambda^m = 1$ and, by the above step, it follows that $z_0 \in \mathbb{J}(f^m) = \mathbb{J}(f)$.

(4) In view of Remarks 5.4.3(2), it is evident that Siegel points belong to $\mathbb{F}(f)$. Now consider a Cremer fixed point z_0. According to Theorem 5.4.2, (f^n) is not uniformly bounded in a neighborhood of z_0. Since $f^n(z_0) = z_0$ for all $n \in \mathbb{N}$, we can conclude that (f^n) is not normal at z_0. □

The following result is a useful sufficient condition for an invariant set $M \subset \mathbb{C}$ to be contained in $\mathbb{J}(f)$ (Exercise 2).

Theorem 6.2.3. *Let f be an entire function and let $M \subset \mathbb{C}$ be an invariant set such that $M \subset \mathbb{B}(f)$. If there exists an $\mu > 1$ satisfying $|f'(z)| \geq \mu$ for all $z \in M$, then M is contained in $\mathbb{J}(f)$.*

In the next theorem, we will use some results about the Fatou components and, therefore, it is convenient to present them prior to Chapter 7.

Theorem 6.2.4. *If f is a transcendental entire function, then $\mathbb{J}(f)$ is infinite and unbounded.*

Proof. Let $g = f^2$. As $\mathbb{J}(f) = \mathbb{J}(g)$, it suffices to prove that $\mathbb{J}(g) \cap D \neq \emptyset$ for all $R > 0$, where $D = \{z \in \mathbb{C} : |z| > R\}$. By Corollary 4.4.4, $g(z) - z$ has infinitely many zeros. Thus, given $R > 0$, D contains two different points a_1 and a_2 satisfying $g(a_j) = a_j$ ($j = 1, 2$). By Picard theorem, ∞ is an essential singularity of g, and we may be sure that there is a $b \in D$ such that $g(b) = a_1$ or $g(b) = a_2$. We will assume that $g(b) = a_2$ and put $D_0 = \{z \in \mathbb{C} : R \leq |z| \leq |a_1| + |a_2| + |b| + 1\}$. To prove by contradiction that $\mathbb{J}(g) \cap D_0 \neq \emptyset$, assume that $D_0 \subset \mathbb{F}(g)$. Then (g^n) is normal in D_0. Since $a_1, a_2 \in D_0 \subset \mathbb{F}(g)$, a_1 and a_2 must be attracting or Siegel points for g. If a_1 is attracting, there exists a disc $\mathbb{D}_r(a_1) \subset D_0$

such that $g^n \to a_1$ uniformly on $\mathbb{D}_r(a_1)$. On the other hand, as (g^n) is normal in D_0, there exists a subsequence $(g^{n_k})_k$ which is uniformly convergent on compact subsets of D_0 to a function h defined on D_0. Obviously, $h_1 \equiv a_1$ in $\mathbb{D}(a_1, r)$ and, in consequence, $h \equiv a_1$ in D_0. But a_2 is a fixed point for g, thus $g^n(a_2) = a_2$ for all $n \in \mathbb{N}$ and, therefore, $h(a_2) = a_2$, which is a contradiction. Then a_1 is a Siegel point for g. By Theorem 7.4.2, there exist a disc $\mathbb{D}_r(a_1) \subset D_0$ and a subsequence $(g^{n_k})_k$ uniformly convergent to z on $\mathbb{D}_r(a_1)$ (notice that the connected component $C(a_1)$ of $\mathbb{F}(g)$ containing a_1 is an invariant component of Fatou, that is, $g(C(a_1)) \subset C(a_1)$). As (g^n) is normal in D_0, there is a subsequence of $(g^{n_k})_k$ uniformly convergent on compact subset of D_0 to a function h analytic in D_0. Then $h(z) \equiv z$ in $\mathbb{D}_r(a_1)$ and, therefore, $h(z) \equiv z$ in D_0. In particular, $h(b) = b$. But we have chosen b such that $g(b) = a_1$, which yields $g^n(b) = a_1$ for all $n \in \mathbb{N}$. In consequence, $h(b) = a_1$ and we again have a contradiction. □

6.3 Fatou exceptional values

For every $z \in \mathbb{C}$, the *backward orbit* of z is defined by

$$O_f^-(z) = \{w \in \mathbb{C} : f^n(w) = z \text{ for some } n \in \mathbb{N}\}.$$

In general, this set is infinite. We will say that z is an *exceptional value* (in the sense of Fatou) for f if $O_f^-(z)$ is a finite set. Note that if z_0 is an exceptional value for f, then the set $f^{-1}(\{z_0\})$ is finite. According to Proposition 3.3.7, transcendental entire functions have at most one exceptional value.

As $O_f^-(z) \subset [z]$, if the grand orbit of z is finite, then z is an exceptional value. Nevertheless, the grand orbit of an exceptional value can be infinite. For rational functions with degree equal to or greater than 2, it is well known that $[z]$ is finite if and only if z is an exceptional value and this value always belongs to the Fatou set; see, for example, [13]. For entire transcendental functions, it is difficult to determine whether it belongs to the Fatou or the Julia set.

Example 6.3.1. Let $E_\lambda(z) = \lambda e^z$. In Chapter 9, we will see that $0 \in \mathbb{F}(E_\lambda)$ for $\lambda \in (0, 1/e)$ and $0 \in \mathbb{J}(E_\lambda)$ for $\lambda > 1/e$.

The next theorem gives a condition under which the exceptional value belongs to the Julia set.

Theorem 6.3.2. *Let f be an entire transcendental function. If $\mathbb{F}(f)$ has no unbounded components, then the exceptional value always belongs to the Julia set.*

Proof. Suppose a is the exceptional value of f and consider the function $g(z) = 1/(f(z) - a)$. Then g is either entire or meromorphic with only a finite number of poles, and we can choose an r_0 such that g is analytic in $\mathbb{C} \setminus \overline{\mathbb{D}}_{r_0}(0)$. As f is entire transcendental, it is easy to prove that $M(r, g) \to \infty$ as $r \to \infty$. Choose $r_1 > r_0$ and let $p_1 = M(r_1, g)$. The set $S_1 = \{z : |g(z)| > p_1\} \cap \mathbb{C} \setminus \overline{\mathbb{D}}_{r_0}(0)$ is nonempty and open. Every z on the boundary of S_1 satisfies $|g(z)| = p_1$. We choose a component of S_1, say D_1. Let us see that D_1 is unbounded. If not, $|g(z)| = p_1$ on its boundary and $|g(z_0)| > p_1$ for some interior point z_0 of D_1. This contradicts the maximum modulus principle. Thus D_1 is unbounded. On the other hand, it is easy to show that g is unbounded in D_1. So, there is a point $b \in D_1$ such that $|g(b)| > p_2 > p_1 + 1$.

We may repeat the argument to determine an open connected $D_2 \subset D_1$ such that $|g(z)| > p_2$ for all $z \in D_2$, with D_2 being unbounded and g unbounded on D_2 (D_2 is a component of the set $S_2 = \{z : |g(z)| > p_2\} \cap \mathbb{C} \setminus \overline{\mathbb{D}}_{r_0}(0)$). Continuing in this way, we get a sequence of domains

$$\cdots \subset D_n \subset D_{n-1} \subset \cdots \subset D_2 \subset D_1$$

such that $|g(z)| > p_k$ on $\partial(D_k)$, D_k is unbounded, and $p_k \to \infty$ as $n \to \infty$. For each $k \in \mathbb{N}$, we choose $w(k) \in \partial(D_k)$ and, therefore, we have $f(w(n)) \to a$ as $n \to \infty$ (see Figure 6.1). Now we connect the points $w(k)$ using a suitable simple path γ such that, along the path, we have $\lim_{z \in \gamma} f(z) = a$.

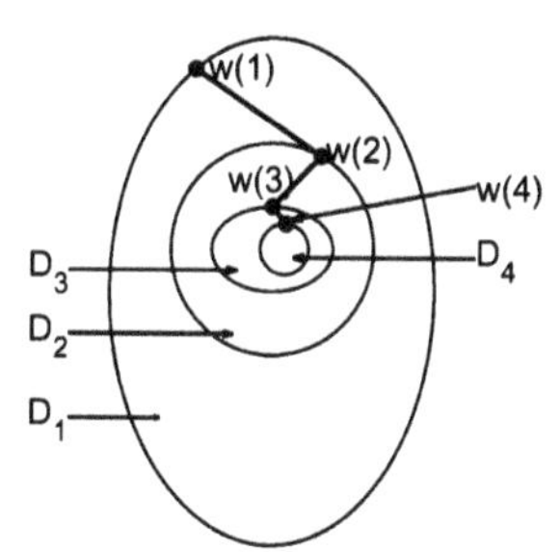

Figure 6.1: The path joining the points $w(k)$'s.

By contradiction, assume that $a \in \mathbb{F}(f)$. As this set is open, there is a neighborhood V of a such that $V \subset \mathbb{F}(f)$. If γ_n denotes the arc of γ that runs from $w(n)$, obviously, there exists an n such that $f(z) \in V$ for all $z \in \gamma_n$ and then $\gamma_n \subset \mathbb{F}(f)$ because $\mathbb{F}(f)$ is completely invariant. Since γ_n is connected, it must be contained in some Fatou component U. Nevertheless, by hypothesis, all components of $\mathbb{F}(f)$ are bounded. This concludes the proof. □

Theorem 6.3.3. *If f is a transcendental entire function, then either $\mathbb{J}(f)$ contains no interior points or $\mathbb{F}(f)$ is empty.*

Proof. Suppose that $\mathbb{J}(f)$ has a nonempty interior. Then there are a $z_0 \in \mathbb{J}(f)$ and a neighborhood U of z_0 contained in $\mathbb{J}(f)$. Since (f^n) is not normal in U, by the two-point condition, it follows that the set $\mathbb{C} \setminus \bigcup_n f^n(U)$ contains at most one point w_0. Since $\mathbb{J}(f)$ is invariant under f, $\bigcup_n f^n(U) \subset \mathbb{J}(f)$. As $\mathbb{F}(f)$ is open, the equality $\mathbb{F}(f) = \{w_0\}$ is impossible. Thus we necessarily have $\mathbb{F}(f) = \emptyset$. □

The following result gives us a characterization of $\mathbb{J}(f)$ in terms of the backwards orbits.

Theorem 6.3.4. *If f is a transcendental entire function, for every nonexceptional point $z_0 \in \mathbb{J}(f)$, we have $\mathbb{J}(f) = \overline{O_f^-(z_0)}$.*

Proof. Given a $z \in \mathbb{J}(f)$ and a neighborhood U of z, as (f^n) is not normal in U, it follows that $\bigcup_n f^n(U)$ intersects $O_f^-(z_0)$ because the latter set is infinite (again we have used the two-point condition). Then there exists an $n \in \mathbb{N}$ such that $f^n(U) \cap O_f^-(z_0) \neq \emptyset$. Choose a $u \in U$ such that $f^n(u) \in O_f^-(z_0)$. Thus $f^m(f^n(u)) = z_0$ for some $m \in \mathbb{N}$. That is, $u \in U \cap O_f^-(z_0)$ and, therefore, $\mathbb{J}(f) \subset \overline{O_f^-(z_0)}$. The converse inclusion is obvious since $\mathbb{J}(f)$ is closed and completely invariant. □

The above theorem may be used as a basis for computing the Julia set (the inverse iteration method). Nevertheless, calculating successive preimages of a point z_0 is a serious problem, unless the function is very simple. In Appendix C, we develop this method applied to the case of the quadratic family $Q_c(z) = z^2 + c$. In Figures C.1, C.2, C.3 and C.4, you can see the Julia set of Q_c for $c = -0.744336 + 0.121198i, -1.3, i$, and 0.25.

Using the same argument, we obtain the following more general result.

Theorem 6.3.5. *If E is a completely invariant set for f containing at least two points, then the closure of E contains $\mathbb{J}(f)$.*

Corollary 6.3.6. *If f is entire, then $\mathbb{J}(f)$ is the smallest set among all completely invariant closed sets with respect to f containing more than two points.*

Recall that a subset A of a topological space X is said to be *perfect* if A does not contain any isolated point.

Theorem 6.3.7. *If f is a transcendental entire function, then $\mathbb{J}(f)$ is perfect.*

Proof. By Theorem 6.2.4, $\mathbb{J}(f)$ is an infinite set. Given a $z \in \mathbb{J}(f)$ and a neighborhood U of z, we have to prove that $(U \setminus \{z\}) \cap \mathbb{J}(f) \neq \emptyset$.

Claim. The set $\mathbb{J}(f) \setminus O_f(z)$ is also infinite. In fact, if $z = f(z)$, then the statement is obvious. Otherwise, if $z \neq f(z)$, then $O_f^-(f(z))$ is necessarily an infinite set. Indeed, as $O_f^-(z) \subset O_f^-(f(z))$, if $f(z)$ is an exceptional value, then such is also z and this contradits Proposition 3.3.7. Thus $O_f^-(f(z))$ is an infinite set contained in $\mathbb{J}(f)$ and we have two possibilities:

(1) $O_f^-(f(z)) \cap O_f(z) = \emptyset$, which implies that the claim is true.
(2) $O_f^-(f(z)) \cap O_f(z) \neq \emptyset$. If w is in the intersection, there exist $n, m \in \mathbb{N}$ such that $f^m(w) = f(z)$ and $f^n(z) = w$. Then $f^{m+n}(z) = f(z)$ and, therefore, the claim holds because $O_f(z)$ is finite.

Hence we can take two different points in $\mathbb{J}(f) \setminus O_f(z)$. By the two-point condition of normality, at least one of these points belongs to $\bigcup_{n\geq 1} f^n(U)$. Thus we have proved that

$$(\mathbb{J}(f) \setminus O_f(z)) \cap \left(\bigcup_{n\geq 1} f^n(U)\right) \neq \emptyset.$$

Finally, for w in the above set, there is an n such that $w = f^n(u)$ with $u \in U$ and $w \in \mathbb{J}(f) \setminus O_f(z)$. Then $u \in U \cap \mathbb{J}(f)$ and $u \neq z$ (note that $u = z$ implies $w = f^n(u) \in O_f(z)$).

In particular, we have proved that $\mathbb{J}(f) \subset \mathbb{J}(f)'$ and, therefore, we have $\mathbb{J}(f) = \mathbb{J}(f)'$ since the converse inclusion is obvious. □

Now we state the so-called first fundamental theorem of the complex dynamics which asserts that $\mathbb{J}(f)$ is the closure of the set formed by all repelling periodic points for f. For rational functions, this result was proved by Fatou. The first proof for entire functions was given by Baker [5]. In his proof, he used the Ahlfors theory of covering surfaces. Here we will see a proof due to Schwick [73] that uses Zalcman's Theorem 2.2.5 for nonnormal families in the unit disc.

Theorem 6.3.8 (First fundamental theorem). *For every transcendental entire function f, $\mathbb{J}(f)$ is the closure of the set formed by all repelling periodic points.*

Proof. Denote by A the set of all completely branched values of f. By Proposition 3.3.7, A contains at most two points.

We fix a point w_0 in $\mathbb{J}(f) \setminus A$ and a disc $D \subset \mathbb{C}/A$ with center w_0. As (f^n) is not normal in D, Zalcman's theorem tells us that there exist a subsequence $(f^{n_k})_k$, complex numbers $(z_k) \to w_0 \in D$, and positive numbers $(\rho_k) \to 0$ such that $h_k(\zeta) = f^{n_k}(z_k + \rho_k\zeta)$ converges to a nonconstant analytic function h uniformly on compact subsets of $\mathbb{C}$. Note that

$$f^{n_k+1}(z_k + \rho_k\zeta) \to (f \circ h)(\zeta). \tag{6.2}$$

Since $w_0 \notin A$, there exist infinite solutions ζ_j of the equation $f(\zeta) - w_0 = 0$ such that $f'(\zeta_j) \neq 0$. On the other hand, if h is transcendental, there is some ζ_j for which the equation $h(\zeta) - \zeta_j = 0$ has a solution α satisfying $h'(\alpha) \neq 0$ (if h is polynomial, this is evident because $h'(\zeta) = 0$ only has a finite number of solutions).

From (6.2), it follows that

$$f^{n_k+1}(z_k + \rho_k\zeta) - (z_k + \rho_k\zeta) \to (f \circ h)(\zeta) - w_0,$$

uniformly on compact subsets of $\mathbb{C}$. As α is a solution of the equation $f \circ h(\zeta) - w_0 = 0$, Hurwiz's theorem asserts that there exists a $k_0 \in \mathbb{N}$ such that, for each $k \geq k_0$, there is a solution α_k of the equation

$$f^{n_k+1}(z_k + \rho_k\zeta) - (z_k + \rho_k\zeta) = 0,$$

with $\alpha_k \to \alpha$. If we put $\overline{z}_k = z_k + \rho_k\alpha_k$, $\overline{z}_k$ is a fixed point of f^{n_k+1} and, in consequence, a periodic point for f. Finally, we will prove that, for k sufficiently large, $|(f^{n_k+1})'(\overline{z}_k)| > 1$ holds. In fact, we have

$$\frac{df^{n_k+1}}{d\zeta}(z_k + \rho_k\zeta)\bigg|_{\zeta=\alpha_k} = \rho_k \frac{df^{n_k+1}}{dz}\bigg|_{z=\overline{z}_k} \to (f \circ h)'(\alpha) \neq 0,$$

since $(f \circ h)'(\alpha) = f'(h(\alpha)) \cdot h'(\alpha) = f'(\zeta_j) \cdot h'(\alpha)$. Thus we have proved that $(f^{n_k+1})'(\overline{z}_k) \to \infty$. Therefore, for large k, $\overline{z}_k$ is a repelling periodic point and $(\overline{z}_k) \to w_0$. The proof concludes because $\mathbb{J}(f)$ has no isolated point. □

The next theorem offers us the following criterion of normality.

Theorem 6.3.9. *The sequence (f^n) is normal at $z_0 \in \mathbb{C}$ if and only if there exists a subsequence that is normal at z_0.*

Proof. Suppose that $(f^{n_k})_k$ is normal at z_0, yet (f^n) is not normal at z_0. By hypothesis, there is a subsequence $(f^{r_j})_j$ of $(f^{n_k})_k$ that is uniformly convergent on a neighborhood U of z_0 to an analytic function h or ∞. As $z_0 \in \mathbb{J}(f)$, it follows from the above theorem that there exists in U some periodic point for f. Then $(f^{r_j})_j$ must be uniformly convergent on U to an analytic function h. According to Weierstrass theorem, $((f^{r_j})')_j \to h'$ uniformly on U. Again by Theorem 6.3.8, there is a periodic repelling point $z^* \in U$. Let p be the prime period of z^* and $\lambda = (f^p)'(z^*)$. We will prove that $(f^{r_j})'(z^*) \to \infty$, which is a contradiction. In fact, for each $r_j > p$, let $r_j = ps_j + r_j^*$ be the result of integer division of r_j by p with $r_j^* \leq p - 1$. By the chain rule, we have

$$(f^{r_j})'(z^*) = [(f^p)'(f^{r_j^*}(z^*))]^{s_j}(f^{r_j^*})'(z^*) = \lambda^{s_j}(f^{r_j^*})'(z^*).$$

Note that $(f^{r_j^*})'(z^*)$ is one of the nonnull complex numbers $(f^k)'(z^*)$ with $k = 0, 1, \ldots, p-1$ and $(s_j) \to \infty$. Since $|\lambda| = |(f^p)'(z^*)| > 1$, it follows that $(f^{r_j})'(z^*) \to \infty$. □

Now we will show that f is chaotic in the Julia set. Recall the definition of chaos due to Devaney.

Definition 6.3.10. Let (X, d) be a metric space and let $f : X \to X$ a map. We will say that f is chaotic if the following statements hold:

(i) f is transitive, that is, for all pairs of open sets U and V of X, there exists an $n \in \mathbb{N}$ such that $f^n(U) \cap V \neq \emptyset$.
(ii) The set of all periodic points for f is dense in X.
(iii) The function f has sensitive dependence on initial conditions, that is, there exists a constant $\delta > 0$ such that, for all $x \in X$ and any open U containing x, there are $y \in U$ and $n \in \mathbb{N}$ such that

$$d(f^n(x), f^n(y)) > \delta.$$

Remark 6.3.11. In 1992, after the appearance of Devaney's definition, it was proven that conditions (i) and (ii) imply (iii) [12].

Theorem 6.3.12. *If f is a transcendental entire function, then the dynamics system $(f, \mathbb{J}(f))$ is chaotic.*

Proof. The first fundamental theorem asserts that condition (ii) holds. Therefore, we only have to prove that f is transitive in $\mathbb{J}(f)$. Let U and V two open sets satisfying:

$\mathbb{J}(f) \cap U \neq \emptyset$ and $\mathbb{J}(f) \cap V \neq \emptyset$. As $\mathbb{J}(f)$ has no isolated point, V contains at least two distinct points. Since (f^n) is not normal in U, Montel's theorem of two points asserts that $\bigcup_{n\geq 1} f^n(U)$ may not be disjoint from V. □

Often this chaotic behavior of the orbits produces aesthetic computer images (see Figures 9.2, C.11, and C.12).

6.4 The escaping set

We have seen that the set formed by all periodic repelling points is dense in $\mathbb{J}(f)$ and, therefore, the filled Julia set $\mathbb{B}(f)$ is also dense in $\mathbb{J}(f)$. In this section we study the relation between $\mathbb{J}(f)$ and the set of all points whose orbits are divergent.

Definition 6.4.1 (Escaping points). If f is an entire function, the set defined by

$$\mathbb{I}(f) = \left\{ z \in \mathbb{C} : \lim_{n\to\infty} f^n(z) = \infty \right\}$$

is called the escaping set of f.

Notice that $\mathbb{I}(f)$ is completely invariant. The study of the escaping set of a polynomial function f, as we will see in Section 6.5, is relatively easy. There, we will show that $\mathbb{I}(f)$ is nonempty and the equality $\mathbb{J}(f) = \partial(\mathbb{I}(f))$ holds. However, when f is transcendental, it is more laborious to prove that $\mathbb{I}(f)$ is nonempty. Actually, we will need the theory of Wiman and Varilon that we develop briefly next.

Suppose that $f(z) = \sum_{n\geq 0} a_n z^n$ is the Taylor series of f at the origin. For each $r > 0$, $|a_n| r^n \to 0$ as $n \to \infty$, and we may consider the *maximal term* $\mu(r,f)$ given by

$$\mu(r,f) = \max\{|a_n| r^n : n \in \mathbb{N}\}.$$

If n_0 is the greatest n such that $\mu(r,f) = |a_n| r^n$, n_0 is denoted by $\nu(r,f)$ and is called the *central index*. The next proposition lists the basic properties of $\mu(r,f)$ and $\nu(r,f)$.

Proposition 6.4.2. *Under the above conditions, if $R > r$, then the following statements hold:*

(i) *$\mu(r,f)$ and $\nu(r,f)$ are nondecreasing and*

$$\lim_{r\to+\infty} \mu(r,f) = \lim_{r\to+\infty} \mu(r,f) = +\infty.$$

(ii) $\frac{\mu(r,f)}{\mu(R,f)} \geq \left(\frac{r}{R}\right)^{\nu(R,f)}$.

(iii) $\mu(r,f) \leq M(r,f) \leq \mu(r,f)(\nu(r,f) + \frac{R}{R-r})$.

Proof. (ii) If we put $\nu = \nu(R,f)$, then

$$\frac{\mu(r,f)}{\mu(R,f)} = \frac{\mu(r,f) r^\nu}{|a_\nu| r^\nu R^\nu} \geq \left(\frac{r}{R}\right)^{\nu(R,f)}.$$

(iii) The first inequality is a consequence of the Cauchy estimates of the derivatives. To prove the second, we proceed as follows:

$$M(r,f) = \max_{|z|=r}|f(z)| \le \left(\sum_{n=0}^{\nu(R,f)-1} + \sum_{n\ge\nu(R,f)} \right)|a_n|r^n.$$

Obviously, the first sum is less than or equal to $\mu(r,f)\nu(R,f)$. For the second, using (ii), we have

$$\begin{aligned}\sum_{n\ge\nu(R,f)} |a_n|r^n &= \sum_{n\ge\nu(R,f)} |a_n|R^n\left(\frac{r}{R}\right)^{\nu(R,f)}\left(\frac{r}{R}\right)^{n-\nu(R,f)}\\ &\le \sum_{n\ge\nu(R,f)} |a_n|R^n\frac{\mu(r,f)}{\mu(R,f)}\left(\frac{r}{R}\right)^{n-\nu(R,f)}\\ &\le \mu(r,f)\sum_{n\ge\nu(R,f)}\left(\frac{r}{R}\right)^{n-\nu(R,f)}.\end{aligned}$$

Now it is easy to obtain the required result. □

Theorem 6.4.3 (Wiman–Valiron). *If f is a transcendental entire function, for each $r > 0$, we can represent its derivatives $f^{(n)}$ as*

$$f^{(n)}(z) = \nu(r,f)^n\left(\frac{z}{z_0}\right)^{\nu(r,f)}\frac{f(z_0)}{z_0^n}(1+\varepsilon_n(r)),$$

where z_0 verifies

$$|z_0| = r, \quad |f(z_0)| = M(r,f),$$

and z is such that

$$|z - z_0| < \frac{r}{\nu(r,f)^{(1/2)+\eta}} \quad (\eta > 0),$$

with $\lim_{r\to+\infty}\varepsilon_n = 0$ uniformly on $z \in \mathbb{D}(z_0, r\nu(r,f)^{-((1/2)+\eta)})$, for r outside an open set L such that $\int_L dt/t$ is finite (the logarithmic measure of L).

Theorem 6.4.4 (Eremenko, 1989 [36]). *If f is a transcendental entire function, then $\mathbb{I}(f) \neq \emptyset$.*

Proof. We denote $\nu(r,f)$ by $\nu(r)$ and choose a $w(r)$ such that

$$|w(r)| = r, \quad |f(w(r))| = M(r).$$

According to Wiman–Valiron theorem, for $\alpha > (1/2)$ and $|z - w(r)| < r\nu(r)^{-\alpha}$, we have

$$f(z) = \left(\frac{z}{w(r)}\right)^{\nu(r)} f(w(r))(1+\varepsilon_1),$$
$$f'(z) = \nu(r)\left(\frac{z}{w(r)}\right)^{\nu(r)} \frac{f(w(r))}{w(r)}(1+\varepsilon_2), \tag{6.3}$$

where $\varepsilon_i = \varepsilon_i(r, z) \to 0$ uniformly with respect to z as $r \to +\infty$ outside a set L with a finite logarithmic measure. Choose an $r_1 > 0$ sufficiently large such that $\nu(r_1) > 10^4$, $r_1 \notin L$, and

$$M(r) > 4r, \quad |\log(1+\varepsilon_1)| < 1 \quad \text{for } r > r_1, z \in \mathbb{D}(w_1, r_1\nu(r_1)^{-\alpha}).$$

Put $w_1 = w(r_1)$ and $M_1 = |f(w(r_1))|$.

It is evident that if $z \in \mathbb{D}(w_1, r_1\nu(r_1)^{-\alpha})$, then z/w_1 belongs to the disc $\mathbb{D}(1, 0.01)$ and, therefore, we may consider the function $\Phi(z)$ defined on the disc $\mathbb{D}(w_1, r_1\nu(r_1)^{-\alpha})$ by $\Phi(z) = \nu(r_1)\log_p(z/w_1) + \log_p f(w_1)$. Here $\log_p$ denotes the principal logarithm. As $\log_p(1+\varepsilon_1)$ is also defined on the mentioned disc, the function $\Phi(z) + \log_p(1+\varepsilon_1)$ is a determined branch of $\log f(z)$ (recall relation (6.3)).

Consider the set $S(1)^*$ (Figure 6.2) defined by

$$S(1)^* = \left\{z \in \mathbb{C} : \left|\log\left|\frac{z}{w_1}\right|\right| < \frac{5}{\nu(r_1)}, \left|\arg\left(\frac{z}{w_1}\right)\right| < \frac{5}{\nu(r_1)}\right\}. \tag{6.4}$$

Note that if $z \in S(1)^*$, then

$$r_1 e^{-5/\nu(r_1)} < |z| < r_1 e^{5/\nu(r_1)} \tag{6.5}$$

holds. First, we will show that $S(1)^* \subset \mathbb{D}(w_1, r_1\nu(r_1)^{-\alpha})$.

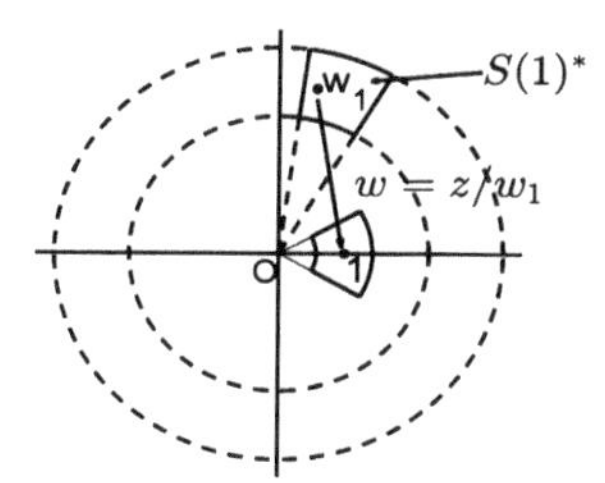

Figure 6.2: The set $S(1)^*$.

Choosing $z \in S(1)^*$, to simplify, we denote z/w_1 by w and $\arg(z/w_1)$ by ϕ, and estimate $|z - w_1|$:

$$|z - w_1| = |w_1|\left|\frac{z}{w_1} - 1\right| = r_1|w - 1|.$$

In the case $|w| < 1$, we have $\sin\phi = a$, $\cos\phi = |w| + b$ and, therefore, $|w - 1| = \sqrt{1 + |w|^2 - 2|w|\cos\phi}$ (see Figure 6.3).

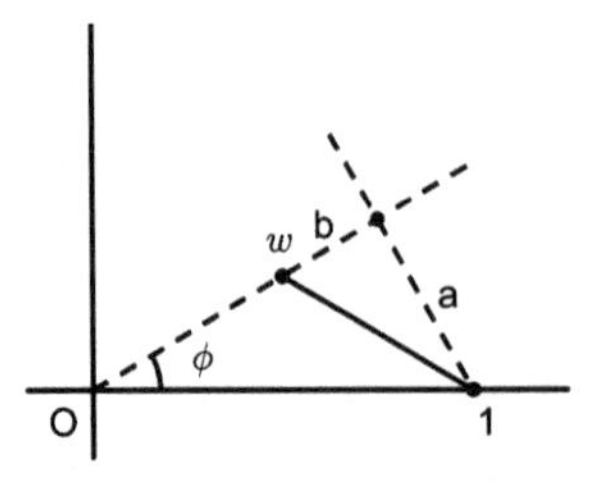

Figure 6.3: The triangle with vertices O, 1, and w.

This equality may be written in the form

$$|w-1| = \sqrt{(1-|w|)^2 - 2|w|(1-\cos\phi)}$$

and then elementary computation yields

$$|w-1| \le (1-|w|) + \sqrt{2|w|}\sqrt{1-\cos\phi} \le (1-|w|) + \frac{5\sqrt{|w|}}{\nu(r_1)}. \tag{6.6}$$

Using the inequalities (6.3), we obtain

$$|w-1| < 1 - e^{-5/\nu(r_1)} + \frac{5e^{5/2\nu(r_1)}}{\nu(r_1)} \le \frac{5}{\nu(r_1)}(1 + e^{5/2\nu(r_1)}). \tag{6.7}$$

For $\nu(r_1) > 10^4$, if we take $\alpha \in (1/2, 1)$, (6.7) implies that $|w-1| < \nu^{-\alpha}$.

The function Φ is one-to-one and maps $S(1)^*$ onto the set Q_1' given by

$$Q_1' = \{\zeta = x + iy : |x - \log M(r_1)| < 5, |y - \arg(f(w_1))| < 5\}.$$

To prove that Φ is onto, for a chosen $\zeta \in Q_1'$, we put

$$z = w_1 e^{(\zeta - \log(f(w_1)))/\nu(r_1)}.$$

It is obvious that $z \in S(1)^*$ and $\Phi(z) = \zeta$.

Now consider the set Q_1 given by

$$Q_1 = \{\zeta = x + iy : |x - \log M(r_1)| < 4, |y - \arg(f(w_1))| < 4\}.$$

For each $w \in Q_1$, we will apply Rouche's theorem in the region $S(1)^*$ to the functions $F(z) = \Phi(z) - w$ and $G(z) = \log(1+\varepsilon_1)$ defined on the disc $\mathbb{D}(w_1, r_1\nu(r_1)^{-\alpha})$, taking as γ the boundary curve of S_1^*. We have to show that $|G(z)| < |F(z)|$ for $z \in \gamma$. As $|\log(1+\varepsilon_1)| < 1$ in the disc, it suffices to prove that $|F(z)| \ge 1$ on γ. The boundary of $S(1)^*$ consists of four similar parts, thus we will do the proof only in the case $\log|(z/w_1)| = 5/\nu(r_1)$. Then

$$|z| = |w_1|e^{5/\nu(r_1)} \quad \text{and} \quad \left|\arg\left(\frac{z}{w_1}\right)\right| \le \frac{5}{\nu(r_1)},$$

which allows us to obtain $\mathbb{R}e(\Phi(z)) = 5 + \log|f(w_1)|$. We proceed to estimate $|\mathbb{F}(z)|$ as follows:

$$\begin{aligned}|F(z)| &\geq |\mathbb{R}e(\Phi(z) - w)| = |5 + \log|f(w_1)| - \mathbb{R}e(w)| \\ &\geq 5 - |\mathbb{R}e(w) - \log|f(w_1)|| \geq 1.\end{aligned}$$

By (6.3), $\log f(z)$ is a bijective map from S_1 onto Q_1. Let us see that $f(S_1)$ contains the set

$$A = \{z \in \mathbb{C} : e^{-4}M(r_1) < |z| < M(r_1)e^4\}.$$

If $\zeta \in A$, then $\log|\zeta| - \log|f(w_1)|| < 4$. If we choose $\arg(\zeta)$ such that $|\arg(\zeta) - \arg(f(w_1))| < \pi$, then $\log\zeta$ belongs to Q_1 and, in consequence, there exists $z \in S_1$ such that $\log z = \log\zeta$. This yields $f(z) = \zeta$.

Next we choose $r_2 \notin L$ such that

$$(1/2)M(r_1) < M(r_2) < 2M(r_1)$$

and put

$$S(2)^* = \left\{z \in \mathbb{C} : \left|\log\left|\frac{z}{w_2}\right|\right| < \frac{5}{v(r_2)}, \left|\arg\left(\frac{z}{w_2}\right)\right| < \frac{5}{v(r_2)}\right\},$$

where $w_2 = w(r_2)$. It is easy to check that this set is contained in A. Repeating the process, we obtain a sequence $(S_j) \to \infty$ such that $S_{j+1} \subset f(S_j)$. The function f is one-to-one in each S_j and, therefore, there exists an analytic branch of f^{-1} defined from S_{j+1} into S_j which we denote by f_j^{-1}. If we put $B_j = f^{-j}(S_{j+1})$, where $f^{-j} = f_1^{-1} \circ f_2^{-1} \circ \cdots \circ f_j^{-1}$, then $\overline{B}_{j+1} \subset B_j$ and this implies that $\bigcap_{j\geq 1} B_j \neq \emptyset$. Finally, taking $z \in \bigcap_{j\geq 1} B_j$, we have $f^j(z) \in S_{j+1}$ for all $j \in \mathbb{N}$, and the proof concludes. □

In the next section, we will prove that the set $\mathbb{I}(f)$ is infinite for all polynomial functions f. However, as mentioned at the beginning of the section, the proof is easier in that case. In the following result, we only use the fact that the escaping set is infinite.

Theorem 6.4.5. *If f is an entire function, then*

$$\partial(\mathbb{I}(f)) = \mathbb{J}(f).$$

Proof. The inclusion $\partial(\mathbb{I}(f)) \subset \mathbb{J}(f)$ follows from a similar argument as that used in the proof of Theorem 6.1.2. Let $z \in \mathbb{J}(f)$ and let U be an open connected neighborhood of z. As $\mathbb{I}(f)$ is an infinite set, applying Montel's theorem to the family (f^n) (which is not normal in U), we deduce that

$$\left[\bigcup_{n\geq 0} f^n(U)\right] \cap \mathbb{I}(f) \neq \emptyset.$$

Since $\mathbb{I}(f)$ is completely invariant, it follows that $U \cap \mathbb{I}(f) \neq \emptyset$ and this proves that $z \in \overline{\mathbb{I}(f)}$. Notice that if the interior of $\mathbb{I}(f)$ is contained in $\mathbb{F}(f)$, then $z \in \partial(\mathbb{I}(f))$. Thus suppose z_0 is an interior point of $\mathbb{I}(f)$ and U is a neighborhood of z_0 contained in $\mathbb{I}(f)$. If $z_0 \in \mathbb{J}(f)$, it follows from the first fundamental theorem that there exists a periodic point $z^* \in U$. The orbit of z^* is bounded, and this yields a contradiction. □

6.5 Polynomial functions

Most of the dynamical properties that we have considered in the previous sections are also valid for polynomial functions. Nevertheless, ∞ is an essential singularity for entire transcendental functions, while it is an attracting fixed point for polynomial functions and, therefore, ∞ belongs to the Fatou set of all polynomials (of degree $d \geq 2$). That is why to study the dynamics of a polynomial, we must consider it as a function from $\hat{\mathbb{C}}$ to itself. Anyway, we will not have this problem because, in this section, we only study some specific properties of the polynomials in relation to the Julia set. In particular, we consider the quadratic family.

Our first result proves that ∞ is an attracting fixed point for all polynomial functions of degree $d \geq 2$ and gives us a method to determine a neighborhood of ∞ contained in its attractive basin.

Theorem 6.5.1. *Let $f(z) = a_0 z^k + a_1 z^{k-1} + \cdots + a_{k-1} z + a_k$ be a polynomial of degree $k \geq 2$. The following statements hold:*

(i) *If $a_0 = 1$, there exists an $R(f) \geq 2$ such that*

$$|f^n(z)| > |f^{n-1}(z)| \left(\frac{|z|^{(k-1)}}{2} \right), \tag{6.8}$$

for all $|z| > R(f)$ and $n \in \mathbb{N}$. In consequence, $\lim_{n\to\infty} f^n(z) = \infty$ uniformly on compact subsets of $|z| > R(f)$ and $|z| < |f(z)| < \cdots < |f^n(z)| < \cdots$ for $n \in \mathbb{N}$ and $|z| > R(f)$.

(ii) *There exists a constant $R(f) > 0$ such that $\lim_{n\to\infty} f^n(z) = \infty$ uniformly on compact subsets of $|z| > R(f)$.*

Proof. (i) From the equality

$$\frac{f(z)}{z^k} = 1 + \frac{a_1}{z} + \cdots + \frac{a_k}{z^k},$$

it follows that

$$\left| \frac{f(z)}{z^k} \right| \geq \left| 1 - \left| \frac{a_1}{z} + \cdots + \frac{a_k}{z^k} \right| \right|. \tag{6.9}$$

As $\lim_{z\to\infty}(a_1/z + \cdots + a_k/z^k) = 0$, we may choose an $R > 2$ such that $|a_1/z + \cdots + a_k/z^k| < 1/2$ for all $|z| > R$. If we put $R(f) = \max\{2, R\}$, this, together with (6.9), allows us to obtain

$$|f(z)| > \frac{|z|^k}{2} = |z|\left(\frac{|z|^{k-1}}{2}\right), \tag{6.10}$$

for all $|z| > R(f)$. Notice that $|z|^{k-1}/2 > 1$ if $|z| > R(f)$, so, applying (6.10) to $f(z)$, we deduce that

$$|f^2(z)| > |f(z)|\left(\frac{|z|^{(k-1)}}{2}\right) > |f(z)|, \quad \text{for all } |z| > R(f) \text{ and } n \in \mathbb{N}.$$

Finally, by induction, it is easy to prove (6.8).

(ii) Now let $f(z) = a_0 z^k + a_1 z^{k-1} + \cdots + a_{k-1} z + a_k$, with $a_0 \neq 1$. If α is a $(k-1)$th root of a_0, the conjugate of f through $\phi(z) = \alpha z$ is the monic polynomial

$$g(w) = \phi \circ f \circ \phi^{-1}(w) = \alpha f\left(\frac{w}{\alpha}\right) = w^k + \alpha a_1 \left(\frac{w}{\alpha}\right)^{k-1} + \cdots + \alpha a_k.$$

Put $R(f) = R(g)/|\alpha|$, where we choose $R(g)$ as in (i). Since $f^n = \phi^{-1} \circ g^n \circ \phi$, statement (ii) follows by a standard argument. □

If $\lim_{n\to\infty} f^n(z) = \infty$ for all $|z| > R$, R is called an escape radius for the polynomial f (for an explicit expression of the escape radius, see Exercise 8).

Theorem 6.5.1 asserts that the set $\{z \in \mathbb{C} : |z| > R(f)\}$ is contained in the attractive basin of the point ∞. As the basin is contained in the Fatou set, it follows that $\mathbb{J}(f) \subset \overline{\mathbb{D}}_{R(f)}(0)$. In fact, we have $\mathbb{B}(f) \subset \overline{\mathbb{D}}_{R(f)}(0)$. The following result is more precise and gives a strategy to obtain a computer image of $\mathbb{B}(f)$ that we will use in Appendix C.

Theorem 6.5.2. *Under the above conditions,*

$$\mathbb{B}(f) = \{z \in \mathbb{C} : |f^n(z)| \le R(f)\ (\forall n \in \mathbb{N})\}.$$

Proof. We only need to prove the inclusion

$$\mathbb{B}(f) \subset \{z \in \mathbb{C} : |f^n(z)| \le R(f)\ (\forall n \in \mathbb{N})\}.$$

Otherwise, let $z \in \mathbb{B}(f)$ be such that there exists an $m \in \mathbb{N}$ satisfying $|f^m(z)| > R(f)$. If we put $w = f^m(z)$, from Theorem 6.5.1, it follows that

$$\infty = \lim_{n\to\infty} f^n(w) = \lim_{n\to\infty} f^{n+m}(z),$$

which is a contradiction. □

Example 6.5.3. If $f(z) = z^4 + z^2 + c$, then $R(f) = \max\{2, \sqrt[4]{4|c|}\}$. In fact, as in the proof of Theorem 6.5.1, we start with the equality

$$\frac{f(z)}{z^4} = 1 + \frac{1}{z^2} + \frac{c}{z^4}.$$

Note that taking z such that $1/|z|^2 < 1/4$ and $|c/z^4| < 1/4$, we have $|\frac{1}{z^2} + \frac{c}{z^4}| < 1/2$. Then, if we put $R(f) = \max\{2, \sqrt[4]{4|c|}\}$,

$$|z| > R(f) \quad \Longrightarrow \quad \left|\frac{1}{z^2} + \frac{c}{z^4}\right| < 1/2.$$

If f is a polynomial function, the escaping set $\mathbb{I}(f)$ is unbounded since $V(f) = \{z \in \mathbb{C} : |z| > R(f)\} \subset \mathbb{I}(f)$. Let us see that $\mathbb{B}(f)$ is the complement of $\mathbb{I}(f)$. It suffices to show that any unbounded orbit is divergent. In fact, suppose that $(f^n(z))$ is unbounded, then there exists an $N \in \mathbb{N}$ such that $|f^N(z)| > R(f)$. Then the orbit of $f^N(z)$ is divergent and, therefore, the orbit of z is divergent, too. Thus we have the equality

$$\partial(\mathbb{B}(f)) = \partial(\mathbb{I}(f)).$$

At the beginning of the chapter, we have seen that, for a general f, the inclusion $\partial(\mathbb{B}(f)) \subset \mathbb{J}(f)$ holds. Now we shall prove that for polynomial functions we have an equality.

Theorem 6.5.4. *If f is a polynomial with degree equal to or greater than 2, then*

$$\partial(\mathbb{B}(f)) = \partial(\mathbb{I}(f)) = \mathbb{J}(f).$$

Proof. Let $z \in \mathbb{J}(f)$ and let U be an open connected neighborhood of z. As $\mathbb{I}(f)$ is an infinite set, Montel's theorem applied to the family (f^n) (which is not normal in U) yields

$$\left[\bigcup_{n\geq 0} f^n(U)\right] \cap \mathbb{I}(f) \neq \emptyset.$$

Since $\mathbb{I}(f)$ is completely invariant, it follows that $U \cap \mathbb{I}(f) \neq \emptyset$ and, therefore, z belongs to $\overline{\mathbb{I}(f)}$. Finally, as the interior of $\mathbb{I}(f)$ is contained in $\mathbb{F}(f)$, we deduce that $z \in \partial(\mathbb{I}(f))$. □

Theorem 6.5.5. *If f is a polynomial of degree equal to or greater than 2, then $\mathbb{B}(f)$ is compact and nonempty, and the set $\mathbb{I}(f)$ is connected.*

Proof. As the degree of f is not less than 2, the equation $f(z) = z$ always has a solution z^* that, obviously, belongs to $\mathbb{B}(f)$. Therefore, it suffices to prove that $\mathbb{I}(f)$ is open. Put $V = \{z : |z| > R(f)\}$ and note that $f(V) \subset V$. Thus, if $z_0 \in \mathbb{I}(f)$, $(f^n(z_0))$ is divergent and there is an $N \in \mathbb{N}$ such that $|f^N(z_0)| > R(f)$. Then $f^N(z_0) \in V$ and this proves the inclusion $\mathbb{I}(f) \subset \bigcup_{n\geq 1}(f^n)^{-1}(V)$. On the other hand, if $z \in (f^n)^{-1}(V)$, then $f^n(z) \in V$ and we conclude that $\lim_{k\to\infty} f^{n+k}(z) = \infty$. We have obtained the equality

$$\mathbb{I}(f) = \bigcup_{n\geq 1} (f^n)^{-1}(V). \tag{6.11}$$

In particular, (6.11) proves that $\mathbb{I}(f)$ is open and, therefore, $\mathbb{B}(f)$ is compact. Now we claim that each $(f^n)^{-1}(V)$ is connected. By contradiction, suppose that the latter set is not connected and note that Theorem 6.5.1 asserts that $V \subset (f^n)^{-1}(V)$. Then there exists a connected component A of $(f^n)^{-1}(V)$ disjoint from V. Thus A is an open subset of $\overline{\mathbb{D}}_{R(f)}$. By the maximum modulus principle, $M = \max\{|f^n(a)| : a \in \overline{A}\}$ is attained at a point z_0 belonging to $\partial(A)$. As $A \subset (f^n)^{-1}(V)$, it follows that $z_0 \in (f^n)^{-1}(V)$. Finally, take a disc $\mathbb{D}(z_0)$ sufficiently small such that it is contained in $(f^n)^{-1}(V)$. Therefore, $A \cup \mathbb{D}(z_0)$ is a connected set contained in $(f^n)^{-1}(V)$ and this contradicts the maximum principle. As we have proved that $(f^n)^{-1}(V)$ is connected, in view of the equality (6.11), the proof concludes because the sequence $((f^n)^{-1}(V))_n$ is nondecreasing. □

We finish this section showing the filled Julia sets of two polynomial functions (Figures 6.4 and 6.5).

Figure 6.4: Filled Julia set (red) of $P(z) = z^4 + 0.5 + 0.66i$.

Figure 6.5: Filled Julia set (red) of $Q(z) = z^2 - 0.7835 + 0.125i$.

6.6 The quadratic family

The quadratic family $Q_c(z) = z^2 + c$, for c a complex constant, is the most important case because its simplicity allows us to apply easily the above results. We denote by $\mathbb{J}_c$ the Julia set of Q_c.

The next result is a simple application of Theorem 6.5.1.

Theorem 6.6.1. *Let $c \in \mathbb{C}$. If $|z| > \max\{\sqrt{2|c|}, 2\}$, then $\lim_{n\to\infty} |Q_c^n(z)| = \infty$.*

Proof. As in Theorem 6.5.1, we consider the equality

$$\frac{Q_c(z)}{z^2} = 1 + \frac{c}{z^2}.$$

Note that if $|z| > \sqrt{2|c|}$, then we have $|cz^{-2}| > 1/2$. Thus, if we put $R(c) = \max\{2, \sqrt{2|c|}\}$, then

$$|Q_c(z)| > |z|, \quad \text{for all } |z| > R(c).$$ □

According to Theorem 6.5.5, the Julia set of $Q_c(z) = z^2 + c$ is contained in the disc $\overline{\mathbb{D}}_{R(c)}(0)$, with $R(c) = \max\{\sqrt{2|c|}, 2\}$.

As an application of Theorem 6.5.2 to the particular case of the function $Q_c(z) = z^2+c$, we have the following useful escape criterion.

Theorem 6.6.2 (Escape criterion). *If $z_0 \in \mathbb{C}$ is such that there exists a $k \in \mathbb{N}$ satisfying $|Q_c^k(z_0)| > R(c)$, then z_0 does not belong to $\mathbb{B}(Q_c)$.*

The two previous results will be used in Appendix C to elaborate the Matlab codes that we need to produce computer graphics of the filled Julia set $\mathbb{B}(Q_c)$. By Theorem 6.6.1, this set is contained in the square $[-R(c), R(c)] \times [-R(c), R(c)]$ and the escaping criterion gives us a stop test.

6.7 Exercises

1. Let f be an entire function. Show that:
   (i) If $f(\overline{z}) = \overline{f(z)}$ for all $z \in \mathbb{C}$, then the Julia set of f is symmetric respect to the real axis.
   (ii) If $f(-z) = -f(z)$ for all $z \in \mathbb{C}$ and $g = -f$, prove that $\mathbb{J}(g) = \mathbb{J}(f)$.
2. Let f be an entire function and let $M \subset \mathbb{B}(f)$ be an invariant set satisfying the following condition: there exists a $\mu > 1$ such that $|f'(z)| \geq \mu$ for all $z \in M$. Prove that M is contained in $\mathbb{J}(f)$.
3. The function $f(z) = (1/2)(e^z - 1)$ has an attracting fixed point at the origin. (a) Show that the half-plane $H = \{z : \mathbb{R}e(z) < 1\}$ is invariant under f. (b) Deduce from (a) that H is contained in the immediate attractive basin of zero. (c) Determine the best $a > 0$ such that $\{z : \mathbb{R}e(z) < a\}$ is contained in the mentioned basin.

4. Let f be an entire function and let $M \subset \mathbb{C}$ be an invariant connected set such that $|f'(z)| \le \mu < 1$ for every $z \in M$. If M contains some point with convergent orbit, then $M \subset \mathbb{F}(f)$.
5. Let $f_\mu(z) = \mu z(1 - z)$ be the complex logistic map. Determine an escape radius for f_μ.
6. (Bergweiler) Let g be an entire function and $\pi(z) = e^z$ for all $z \in \mathbb{C}$. Assume that $g \circ \pi(z) \neq 0$ for every $z \in \mathbb{C}$, then there exists an entire function f such that $\pi \circ f = g \circ \pi$ ($g \circ \pi(z) \neq 0$ for every $z \in \mathbb{C}$, so there is an analytic branch of $\log g \circ \pi(z)$ defined on $\mathbb{C}$. In [15], the equality $\mathbb{F}(f) = \pi^{-1}(\mathbb{F}(g))$ is proved whenever f is not a polynomial of degree less than 2.
   (a) Prove the inclusion $\pi^{-1}(\mathbb{F}(g)) \subset \mathbb{F}(f)$.
   (b) Consider the example $f(z) = 2z$ and $g(z) = z^2$.
7. Let $f(z) = 1 + z + e^{-z}$. Show that the rays $z = x + (2k + 1)\pi i$ $(x \le 0)$ are contained in $\mathbb{J}(f)$.
8. (An explicit expression of the escape radius) Let $P(z) = a_n z^n + a_{n-1} z^{n-1} + \cdots + a_1 z + a_0$ be a polynomial of degree $n \ge 2$ with $a_n \neq 0$. Show that $R = (1 + \sum_k |a_k|)/|a_n|$ is an escape radius for P.
9. Let U be a region in $\mathbb{C}$ such that $\mathbb{C} \setminus U$ contains more than two points and let $f : U \to U$ be analytic. If f has an attracting fixed point z^* in U, prove that $f^n(z) \to z^*$ uniformly on compact subsets of U.
10. It is well known that $\cos(kz)$ can be expressed as a polynomial T_k in $\cos z$, namely

$$T_k(\cos z) = \cos(kz).$$

The T_k's are called the Tchebychev polynomials (for example, $T_2(x) = 2x^2 - 1$). Using this relationship, show that:
   (i) $T_k^n(z) \to \infty$ for all $z \notin [-1, 1]$.
   (ii) $T_k^n([-1, 1]) \subset [-1, 1]$.
   (iii) $\mathbb{J}(T_k) = [-1, 1]$.
11. Let $f(z) = z^2 + c$, with $c > 1/4$. Prove that $\mathbb{J}(f)$ is disconnected.
12. Suppose that P is a polynomial of degree $n \ge 2$ and that $\mathbb{J}(P)$ is the unit circle. Prove that $P(z) = az^n$ with $|a| = 1$.

Bibliography remarks

In Section 6.3, the proof of Theorem 6.3.8 is based on [73].

Section 6.4 is based on [63, Section 3.1] and [36].

Section 6.5 is inspired by [63, Section 1.1].

7 Fatou components

The classification theorem was proved by Fatou and Cremer for rational functions. In 1932, Cremer classified the Fatou components with the property that (f^n) has a nonconstant limit function. Combining this with earlier results of Fatou on the stable components in which (f_n) has constant limit functions only, the classification theorem was established. In this chapter, we follow paper [11] by Baker, Kotus, and Lü where they stated the classification theorem for meromorphic functions. Sections 7.2 to 7.4 are based on some personal work and inspired by [47] and [11].

7.1 Fatou components. Classification

Throughout this chapter, f denotes a transcendental entire function. If U is a connected component of the Fatou set $\mathbb{F}(f)$, we say that U is a *Fatou component*. As $\mathbb{F}(f)$ is open, its components are maximal mutually disjoint regions. The number of components may be finite or infinite, but countable. If U is a concrete Fatou component, since $f^n(U)$ is also a region contained in $\mathbb{F}(f)$, it follows that $f^n(U)$ is necessarily contained in a certain Fatou component, say U_n. There are several possibilities for the orbit of U under f:

1) *Wandering domains.* All the U_n's are pairwise disjoint ($U_i \cap U_j = \emptyset$ if $i \neq j$). In this case, we say that U is a wandering domain of $\mathbb{F}(f)$. It is well known that rational functions do not have wandering domains (Sullivan, 1985 [76]). However, in the framework of transcendental entire functions, those domains may exist. Baker, in 1976, showed an example of an entire function with wandering domains.
2) *Periodic components.* If $f^n(U) \subset U$, we will say that U is a periodic component and, if n is the smallest positive integer satisfying the above condition, n is called the period of U. If we put $U_0 = U$, $\{U_0, U_1, \ldots, U_{n-1}\}$ is the n-cycle determined by U. Then f maps each component U_j into U_{j+1}. If $n = 1$, then $f(U) \subset U$ and we say that U is a *fixed, or invariant, component* of $\mathbb{F}(f)$. If, in addition, the condition $f^{-1}(U) \subset U$ holds, then U is called *completely invariant.*
3) *Preperiodic components.* If U_k is periodic for some $k \geq 1$, we will say that U is a preperiodic component of $\mathbb{F}(f)$.

Let U and V be Fatou components of a rational function f. In [13], it is proved that the inclusion $f(U) \subset V$ can only happen if $f(U) = V$. However, for transcendental entire functions, the strict inclusion $f(U) \subset V$ is possible. Next, we see an example of this fact.

Example 7.1.1. Let $f(z) = \lambda e^z$ for $\lambda \in (0, 1/e)$. In Chapter 9, we will see that f has two real fixed points: $q_\lambda < 1$ (attracting) and $p_\lambda > 1$ (repelling). If $\ell_\lambda \in (\log(1/\lambda), p_\lambda)$, Proposition 9.1.3 asserts that the half-plane $\mathbb{R}e(z) < \ell_\lambda$ is invariant under f and contained in $\mathbb{F}(f)$. Denote by U the Fatou component containing the mentioned half-plane and let V_0 be the horizontal strip $\pi/2 < \mathbb{I}m(z) < 3\pi/2$. For $z \in V_0$, we have

https://doi.org/10.1515/9783111689685-007

$$\mathbb{R}e(f(z)) = \lambda e^{\mathbb{R}e(z)} \cos(\mathbb{I}m(z)) < 0$$

and, therefore, $f(V_0) \subset U$. Since $\mathbb{F}(f)$ is completely invariant, V_0 is contained in the Fatou set and we may consider the Fatou component V which contains V_0. It is evident that $U = V$ and $f(U) \subset U$. However, $0 \in U$ but $0 \notin f(U)$.

Herring [44] and Bergweiler–Rhode [14] independently proved the following result. In Chapter 8, we introduce the notion of an asymptotic value.

Theorem 7.1.2. *Let f be a transcendental entire function and let U and V be Fatou components such that $f(U) \subset V$. Then $V \setminus f(U)$ contains at most one point which is an asymptotic value of f.*

The following fundamental theorem proves that the behavior of the iterates of f on the periodic components is completely classified.

Theorem 7.1.3 (Classification of periodic components). *Let f be a transcendental entire function and let U be an m-periodic component of $\mathbb{F}(f)$. Then exactly one of the following cases holds:*

(1) *Attractive component. The set U contains an m-periodic attracting point z^*. Then $f^{mn}(z) \to z^*$ as $n \to \infty$ for all $z \in U$. Thus, U is the immediate attractive basin of z^*.*
(2) *Parabolic component. There exists a fixed point z^* of f^m belonging to $\partial(U)$ with multiplier equal to 1 and with the property that all orbits starting in U converge to z^* (orbits under f^m). Then U is called a Leau, or parabolic, domain.*
(3) *Siegel disc. The set U is simply connected and the restriction of f^m to U is conjugate to an irrational rotation on $\mathbb{D}$.*
(4) *Baker domain. The set U is not bounded in $\mathbb{C}$ and $\lim_{k\to\infty} f^{mk}(z) = \infty$, for all $z \in U$.*

Without loss of generality, we assume that $m = 1$, that is, $f(U) \subset U$.

Theorem 7.1.3 will be proved later. Before, we will develop all the necessary properties about Fatou components and the set formed by all possible limit functions of the sequence (f^n). There is another approach to prove the classification theorem that uses the hyperbolic metric (see, for instance, [63]).

7.2 Multiply connected components

In this section, it is proved that, for transcendental entire functions, every preperiodic Fatou component is simply connected. Recall that, given a closed curve γ and an $a \in \mathbb{C} \setminus \gamma$, the *index of γ with respect to the point a* is defined by

$$n(\gamma, a) = \frac{1}{2\pi i} \oint_{\gamma} \frac{dz}{z - a}.$$

Theorem 7.2.1. *Let f be a transcendental entire function and let Ω be a multiply connected Fatou component. If γ is a Jordan curve in Ω which is not contractible in Ω, then the following statements hold:*
(1) $(f^n) \to \infty$ *uniformly on compact subsets of* Ω.
(2) *There exists an* $n_0 \in \mathbb{N}$ *such that* $n(f^n \circ \gamma, 0) \neq 0$ *for* $n \geq n_0$.

Proof. (1) First, we prove that $(f^n) \to \infty$ uniformly on γ. Otherwise, there exist an $M > 0$, a strictly increasing sequence $(n_k)_k$, and $z_k \in \gamma$ such that

$$|f^{n_k}(z_k)| \leq M \quad \text{for all } k \in \mathbb{N}. \tag{7.1}$$

By the normality of (f^n), it follows from (7.1) that $(f^{n_k})_k$ admits a subsequence $(f^{m_k})_k$ which is uniformly convergent on compact subsets of Ω to a function g defined and analytic on Ω. As γ is compact, there exists a $c > 0$ such that $|g(z)| < c$ for all $z \in \gamma$. Since $(f^{m_k})_k \to g$ uniformly on γ, there is a k_0 such that $|f^{m_k}(z)| \leq c$ for all $z \in \gamma$ and $k \geq k_0$. By hypothesis, there is a point $z_0 \in \mathbb{J}(f)$ in the interior region of γ. Take a disc $\mathbb{D}_r(z_0)$ contained in $\operatorname{int}(\gamma)$. As $\mathbb{J}(f)$ is the closure of the set of periodic repelling points, there exists a p-periodic repelling point $z_1 \in \mathbb{D}_r(z_0)$. If $\mu = (f^p)'(z_1)$, then $|\mu| > 1$.

Applying the Cauchy integral formula, we obtain

$$|(f^{m_k})'(z)| \leq \frac{cL(\gamma)}{2\pi d^2} \quad \text{for } z \in \mathbb{D}_r(z_0),\ k \geq k_0, \tag{7.2}$$

where $d = \min\{|z - w| : z \in \gamma, w \in \overline{\mathbb{D}}_r(z_0)\}$. For each $m > p$, let $m = kp + m_0$ be the result of the integer division of m by p, where $m_0 < p$. Now the chain rule yields

$$|(f^m)'(z_1)| = |\mu|^k |(f^{m_0})'(z_1)|. \tag{7.3}$$

From the equality $(f^p)'(z_1) = (f^{p-n})'(f^n(z_1))(f^n)'(z_1)$, valid for $n = 1, 2, \ldots, p-1$, we deduce that $(f^n)'(z_1) \neq 0$, for $n = 1, 2, \ldots, p-1$. Since $|\mu| > 1$, (7.3) contradicts (7.2). Thus we have proved that $(f^n) \to \infty$ uniformly on γ.

Now the proof of (1) is very easy. Again, arguing by contradiction, suppose that (1) is not true. Then there exist a compact $K \subset \Omega$, $M > 0$, f^{n_k}, and $z_k \in K$ such that

$$|f^{n_k}(z_k)| \leq M \quad \text{for all } k \in \mathbb{N}. \tag{7.4}$$

A subsequence $(f^{m_k})_k$ of $(f^{n_k})_k$ converges uniformly to ∞ on compact subsets of Ω (recall we have proved that $(f^n) \to \infty$ on γ). Obviously, this contradicts (7.4).

(2) Notice that

$$n(f^n \circ \gamma, 0) = \frac{1}{2\pi i} \oint_{f^n \circ \gamma} \frac{dz}{z} = \frac{1}{2\pi i} \oint_{\gamma} \frac{(f^n)'(z)dz}{f^n(z)}.$$

If (2) is false, then there exists a strictly increasing sequence (n_k) such that $n(f^{n_k} \circ \gamma, 0) = 0$ for all $k \in \mathbb{N}$. According to the argument principle, f^{n_k} has no zeros in the interior of γ. Applying the maximum modulus principle to each $1/|f^{n_k}|$, we deduce that the absolute minimum of $|f^{n_k}|$ in $\gamma \cup \operatorname{int}(\gamma)$ is attained in γ. Let z_k be the point where $|f^{n_k}|$ attains its minimum. By (1), we know that $f^{n_k} \to \infty$ on γ and, therefore, also on $\operatorname{int}(\gamma)$ because $|f^{n_k}(z)| \geq |f^{n_k}(z_k)|$ for all $z \in \gamma \cup \operatorname{int}(\gamma)$. This is not possible since there are periodic points for f in $\operatorname{int}(\gamma)$. □

Theorem 7.2.2. *Let f be a transcendental entire function that is bounded on a curve Γ going to infinity. All Fatou components of $\mathbb{F}(f)$ are simply connected.*

Proof. Suppose that there exists a Fatou component Ω which is multiply connected and consider a Jordan curve γ in Ω containing in its interior some point of $\mathbb{J}(f)$. By Theorem 7.2.1, there is an $n_0 \in \mathbb{N}$ such that $n(f^n \circ \gamma, 0) \neq 0$ for $n \geq n_0$. The $f^n \circ \gamma$'s are closed curves whose distances to the origin grow. Then, for n sufficiently large, each $f^n \circ \gamma$ intersects Γ. Let $z_n \in \Gamma \cap (f^n \circ \gamma)$. Obviously, $f(z_n) \in f^{n+1} \circ \gamma$ and, therefore, $f(z_n) \to \infty$ which contradicts the boundedness of f on Γ. □

We need the following technical lemma.

Lemma 7.2.3. *Let f be a transcendental entire function and let γ be a Jordan curve such that $(f^n) \to \infty$ uniformly on γ and $n(f^n \circ \gamma, 0) \neq 0$ for all $n \in \mathbb{N}$. For every positive constant $c > 0$, there exists an $n_0 \in \mathbb{N}$ such that, for each $n \geq n_0$, there is a $z \in f^n \circ \gamma$ satisfying $|f(z)| > |z|^c$.*

Proof. By contradiction, suppose that there exists a strictly increasing sequence of positive integers $(n_k)_k$ such that

$$|f(z)| \leq |z|^c \quad \text{for all } z \in f^{n_k} \circ \gamma \text{ and } k \in \mathbb{N}.$$

Choose $N \in \mathbb{N}$ such that $N > c$ and

$$|f(z)| \leq |z|^N \quad \text{for all } z \in f^{n_k} \circ \gamma \text{ and } k \in \mathbb{N}. \tag{7.5}$$

To simplify, we denote by S_k the trajectory of the arc $f^{n_k} \circ \gamma$. We construct inductively sequences of positive numbers $r(k) \to +\infty$ and a strictly increasing sequence of positive integers $(m_k)_k$ such that

$$\min\{|z| : z \in S_{m_k}\} < r(k) < \max\{|z| : z \in S_{m_{k+1}}\}. \tag{7.6}$$

Put $m_1 = 1$ and choose $r(1) > 0$ such that $r(1) > \max\{|z| : z \in S_{m_1}\}$. Since $\lim_{n\to\infty} f^n \circ \gamma = \infty$, we can take $m_2 > m_1$ such that $r(1) < \min\{|z| : z \in S_{m_2}\}$. Next, we choose $r(2) > 1 + \max\{|z| : z \in S_{m_2}\}$ and $m_3 > m_2$ such that $r(2) < \min\{|z| : z \in S_{m_3}\}$. Let $G(k)$ be the connected component (bounded) of $\mathbb{C}/S_{m_k}$ containing the origin (recall that $n(f^n \circ \gamma, 0) \neq 0$ for $n \in \mathbb{N}$). By construction, $C(0, r(k)) \subset G(k+1) \setminus \overline{G(k)}$ (see Figure 7.1).

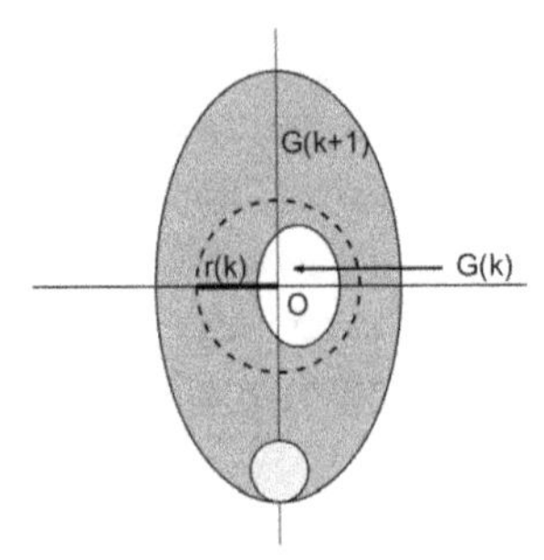

Figure 7.1: The sets $G(k)$, $G(k+1)$, and the circle of radius $r(k)$.

If we denote by L_k the connected component of $G(k+1) \setminus \overline{G(k)}$ that contains $C(0, r(k))$, then

$$\partial(L_k) \subset \partial(G(k+1)) \cup \partial(G(k)) \subset S_{m_k} \cup S_{m_{k+1}}.$$

Thus $|f(z)| \leq |z|^N$ for all $z \in L_k$ and $k \in \mathbb{N}$. As $0 \notin L_k$, it follows from the maximum modulus principle that

$$\left|\frac{f(z)}{z^N}\right| \leq 1 \quad \text{for all } z \in L_k,\ k \in \mathbb{N}.$$

In particular, we have

$$|f(z)| \leq |z|^N \quad \text{for all } |z| = r(k) \text{ and } k \in \mathbb{N}.$$

Therefore, we conclude that $M(r(k)) \leq r(k)^N$ for $k \in \mathbb{N}$. However, this is a contradiction because

$$\lim_{r \to +\infty} \frac{\log M(r)}{\log r} = \infty$$

for all transcendental entire functions. □

Theorem 7.2.4 (Baker). *If f is a transcendental entire function, then every unbounded Fatou component is simply connected.*

Proof. Suppose that Ω is an unbounded Fatou component that is not simply connected. Then we can choose a Jordan curve γ in Ω containing in its interior points of $\mathbb{J}(f)$. According to Theorem 7.2.1, $n(0, f \circ \gamma) \neq 0$ for $n \geq n_0$ and $(f^n) \to \infty$ uniformly on γ. Then, for n sufficiently large, $f^n \circ \gamma$ intersects Ω. Thus, for $n \geq n_0$, $\Omega \cup (f^n \circ \gamma)$ is a connected set contained in $\mathbb{F}(f)$ and, by the maximality of Ω, the curves $f^n \circ \gamma$ are in Ω for $n \geq n_0$. Let $\gamma_0 = f^{n_0} \circ \gamma$ and $\gamma_1 = f \circ \gamma_0$. Choose a bounded region Ω_0 containing γ_i for $i = 0, 1$ and such that $\overline{\Omega_0} \subset \Omega$. Again by Theorem 7.2.1, there exists an $m_0 \in \mathbb{N}$ such that $|f^n(z)| > 1$ for $z \in \overline{\Omega_0}$ and $n \geq m_0$. Now we apply the Harnack inequality to the functions $h_n(z) = \log |f_n(z)|$ and deduce that

$$h_n(w) \leq c h_n(z) \quad \text{for } w \in \gamma_1, z \in \gamma_0, n \geq m_0,$$

where c is a certain positive constant depending on Ω and $\overline{\Omega_0}$. Equivalently,

$$|f^n(w)| \leq |f^n(z)|^c \quad \text{for } w \in \gamma_1, z \in \gamma_0, n \geq m_0. \tag{7.7}$$

On the other hand, by the above lemma, there exists a $\zeta \in f^n \circ \gamma$ for n sufficiently large such that

$$|f(\zeta)| > |\zeta|^c. \tag{7.8}$$

Taking n such that $n - n_0 > m_0$, choose a $z_0^* \in \gamma$ satisfying $\zeta = f^n(z_0^*)$. If we put $z_0 = f^{n_0}(z_0^*)$, then $z_0 \in \gamma_0$, $\zeta = f^{n-n_0}(z_0)$, and $z_1 = f(z_0) \in \gamma_1$. Finally, in view of (7.8), we deduce that

$$|f(\zeta)| = |f^{n-n_0}(z_1)| > |\zeta|^c = |f^{n-n_0}(z_0)|^c,$$

which contradicts (7.7). □

Corollary 7.2.5. *Let f be a transcendental entire function. Every preperiodic Fatou component is simply connected and every multiply connected component is bounded and wandering. Moreover, if $\mathbb{F}(f)$ has an unbounded component D, then every Fatou component is simply connected.*

Proof. (1) Let Ω be a preperiodic component. If it is unbounded, then Theorem 7.2.4 asserts that Ω is simply connected. If Ω is bounded and preperiodic, there exists a k such that $f^k(\Omega)$ is m-periodic. Thus $f^{m+k}(\Omega) \subset f^k(\Omega)$. As $f^k(\Omega)$ is bounded, it follows that some sequence $(f^{n_k})_k$ is uniformly bounded in Ω. According to Theorem 7.2.1, Ω cannot be multiply connected.

(2) By (1), if Ω is multiply connected, then it is not preperiodic and Theorem 7.2.4 tells us that it must be bounded.

(3) Suppose that Ω is a multiply connected component. By Theorem 7.2.1, $(f^n) \to \infty$ uniformly on compact subsets of Ω. Take a Jordan curve γ in Ω such that is not contractible in Ω. The $f^n \circ \gamma$'s are closed curves whose distances to the origin grow. As D is unbounded, for n sufficiently large, $f^n \circ \gamma$ intersects D. Then $f^n(\Omega) \cap D \neq \emptyset$ for $n \geq n_0$ and, in consequence, $f^n(\Omega) \subset D$ for $n \geq n_0$. Thus Ω is not a wandering domain and this contradicts (2). □

The above result proves that transcendental entire functions have no Herman rings. An m-periodic Fatou component Ω of $\mathbb{F}(f)$ is called a *Herman ring* if it is doubly connected and f^m is conformally conjugate to a rotation on an annulus $\{z \in \mathbb{C} : 1 < |z| < r\}$. Polynomial functions have no Herman rings either (Exercise 7). The first examples of Herman rings were produced by Sullivan in 1984.

We say that X is *totally disconnected* if every connected subset of X is a singleton.

There are polynomials whose Julia set is totally disconnected. However, this does not occur in the framework of transcendental entire functions.

Corollary 7.2.6. *If f is a transcendental entire function, then $\mathbb{J}(f)$ is not totally disconnected.*

Proof. The complement of a totally disconnected set is unbounded and multiply connected. □

7.3 Limit functions

A function ϕ is a *limit function* of (f^n) in a Fatou component U if there exists some subsequence of (f^n) which is uniformly convergent on compact subsets of U to ϕ. We will denote by $\mathbb{L}(U)$ the set of all limit functions (notice that, if some subsequence of (f^n) diverges uniformly on compact subsets of U, then the constant function $\phi \equiv \infty$ belongs to $\mathbb{L}(U)$).

In order to prove the classification theorem, we need to study the properties of this set.

Lemma 7.3.1. *Let U be a fixed Fatou component. If there exists a constant limit function $z_0 \in \hat{\mathbb{C}}$, then z_0 is a fixed point of f or $z_0 = \infty$.*

Proof. Let $z_0 \neq \infty$ be such that

$$z_0 = \lim_{k\to\infty} f^{n_k}(z) \quad \text{for all } z \in U.$$

Applying f to each side of the equality, we obtain

$$f(z_0) = \lim_{k\to\infty} f(f^{n_k}(z)) = \lim_{k\to\infty} f^{n_k}(f(z)) = z_0$$

since $f(z) \in U$. □

Lemma 7.3.2. *Let U be a Fatou component. Then $\mathbb{L}(U)$ does not contain any repelling fixed point of f.*

Proof. Suppose that there exists a repelling fixed point $a \in \mathbb{L}(U)$ for f. Note that $f^n(z) \neq a$ for all $z \in U$ and $n \in \mathbb{N}$ since $a \in \mathbb{J}(f)$ and $f^n(z) \in \mathbb{F}(f)$. As $a \in \mathbb{L}(U)$, there exists a subsequence $(f^{n_k})_k$ of (f^n) such that $a = \lim_{k\to\infty} f^{n_k}(z)$ for all $z \in U$. According to Remarks 4.1.4, it follows that, for each $z \in U$, there is a $k_0 \in \mathbb{N}$ such that $f^{n_k}(z) = a$ for all $k \geq k_0$. So, we have reached a contradiction. □

Theorem 7.3.3. *Suppose that U is a fixed Fatou component and every limit function in $\mathbb{L}(U)$ is constant. Then $\mathbb{L}(U)$ contains only an element b and $f^n(z)$ converges to b uniformly on compact subsets of U.*

Proof. First, we prove that $\mathbb{L}(U)$ is a singleton. If not, $\mathbb{L}(U)$ contains at least two values a and b. One of them will be finite, say b. By Lemma 7.3.1, b is a fixed point of f. Since the zeros of $f(z) - z$ are isolated points, we may choose an $\epsilon > 0$ such that $\overline{\mathbb{D}}_c(b, \epsilon)$ does not contain any point of $\mathbb{L}(U)$, except b ($\overline{\mathbb{D}}_c(b, \epsilon)$ denotes the closed ball of radius ϵ with center b in $(\hat{\mathbb{C}}, d_c)$). Fix $z_0 \in U$ and consider the set

$$M = \{m \in \mathbb{N} : d_c(f^m(z_0), b) < \epsilon\}.$$

Inductively, we shall construct a strictly nondecreasing sequence $(m_j)_j$ of positive integers satisfying $m_j \notin M$ and $m_j - 1 \in M$. Let m_1 be the first positive integer m greater than 1 such that $m \notin M$. There exist infinitely many m satisfying this property because there is a subsequence $(f^{n_k}(z))_k$ that converges uniformly on compact subsets of U to $a \neq b$ (we take the m_j's from the subsequence $(n_k)_k$). Note that if $n_k \in M$ for $k \geq k_0$, then $d_c(a, b) \leq \epsilon$, which is not possible. By construction of m_1, the integer $m_1 - 1$ belongs to M. Next, let m_2 be the first positive integer greater than m_1 so that it is not in M and, therefore, $m_2 - 1 \in M$. Proceeding in this way, we get a sequence $(m_j)_j$ satisfying $f^{m_j}(z_0) \to a$. Considering a subsequence, if necessary, $f^{m_j - 1}(z_0) \to c \in \mathbb{L}(U)$ because (f^n) is normal in U. Then $d_c(c, b) \leq \epsilon$ and this is only possible if $c = b$. Finally, from the equality

$$f^{m_j}(z_0) = f(f^{m_j - 1}(z_0)),$$

the contradiction follows:

$$a = f(b) = b.$$

Now to conclude the proof, we need to show that $f^n(z) \to b$ uniformly on compact subsets of U. Reasoning by contradiction and using the chordal distance (because b may be ∞), it is a simple exercise. □

Theorem 7.3.4. *Under the conditions of Theorem 7.3.3, suppose that b is the only limit function that $\mathbb{L}(U)$ contains. Then exactly one of the following statements holds:*
(1) *One has $b = \infty$.*
(2) *Point b is an attracting fixed point of f and $b \in U$.*
(3) *One has $f'(b) = 1$ and $b \in \partial(U)$.*

Proof. We suppose that $b \neq \infty$. The above theorem tells us that $f^n(z) \to b$ for all $z \in U$, therefore, we have $b \in \overline{U}$. From the previous lemmas, we know that b is a fixed point for f and is not repelling. There are two possible cases:
(2) $b \in U$. As $f^n(z) \to b$ uniformly on compact subsets of U, Weierstrass theorem asserts that $(f^n)'(b) \to 0$. Then there exists an $n \in \mathbb{N}$ such that $|(f^n)'(b)| < 1$. Since $(f^n)'(b) = f'(b)^n$, it follows that $|f'(b)| < 1$.

(3) $b \in \partial(U)$. We shall prove that $|f'(b)| = 1$. Since each attracting fixed point belongs to the interior of a Fatou component, b cannot be attracting. On the other hand, by Lemma 7.3.2, b cannot be repelling. Then necessarily $|f'(b)| = 1$.
Let us see that really $f'(b) = 1$. To simplify, we suppose that $b = 0$. Thus $f(0) = 0$ and $|f'(0)| = 1$. As $f'(0) \neq 0$, there exists a disc $\mathbb{D}_\epsilon$ where f is univalent. Choose a $\delta \in (0, \epsilon)$ such that $|f(z)| < \epsilon$ for $|z| < \delta$. Given $z_0 \in \mathbb{D}_\delta$, put $z_1 = f(z_0) \in \mathbb{D}_\epsilon$. Let V be a relatively compact region such that $\overline{V} \subset U \cap \mathbb{D}_\epsilon$ and $z_0, z_1 \in V$. Since $f^n(z) \to 0$ uniformly on V, there exists an n_0 satisfying $|f^n(z)| < \epsilon$ for all $z \in V$ and $n \geq n_0$. The set $V_0 = \bigcup_{n \geq n_0} f^n(V)$ is invariant for f and f^n is univalent in V_0 for all $n \in \mathbb{N}$ (note that $V_0 \subset U \cap \mathbb{D}_\epsilon$). Furthermore, V_0 is connected. Indeed, put $\zeta_1^n = f^n(z_1)$ for each $n \geq n_0$ and note that $\zeta_1^n \in f^n(V) \cap f^{n+1}(V)$. Now having chosen $\zeta_0 \in V_0$, consider the functions $\phi_n(z) = f^n(z)/f^n(\zeta_0)$ defined in V_0. Notice that

$$f^n(z) \neq 0 \quad \text{for all } n \in \mathbb{N} \text{ and } z \in U, \tag{7.9}$$

since $f^n(z)$ belongs to U and $0 \in \partial(U)$. Those functions are such that:
(i) They are univalent in V_0 since every f^n is.
(ii) One has $\phi_n(\zeta_0) = 1$ for all $n \in \mathbb{N}$.
(iii) Also $0 \notin \phi_n(V_0)$ for all $n \in \mathbb{N}$, for the same reason as in (7.9).
Since the functions ϕ_n do not take the values 0 and 1 in $V_0 \setminus \{\zeta_0\}$, it follows that (ϕ_n) is normal in $V_0 \setminus \{\zeta_0\}$.

Claim. (ϕ_n) is normal in V_0.

Given a subsequence $(\phi_{n_k})_k$, there exists a subsequence $(\phi_{m_k})_k$ uniformly convergent on compact subsets of $V_0 \setminus \{\zeta_0\}$. First, we prove that it necessarily converges to a function ϕ defined and analytic in $V_0 \setminus \{\zeta_0\}$. In fact, if the convergence is to ∞, let $r > 0$ be such that $\overline{\mathbb{D}}_r(\zeta_0) \subset V_0$ and denote by γ its boundary. By (iii) and the maximum modulus principle, each $|\phi_{m_k}|$ attains its absolute minimum in $\overline{\mathbb{D}}_r(\zeta_0)$, in a point belonging to γ. There exists a $k_0 \in \mathbb{N}$ such that $|\phi_{m_k}(z)| > 1$ for $k \geq k_0$ and $z \in \gamma$. Then $|\phi_{m_k}(\zeta_0)| > 1$ for $k \geq k_0$, which contradicts (ii). Therefore, $(\phi_{m_k})_k$ converges uniformly on compact subsets of $V_0 \setminus \{\zeta_0\}$ to a function ϕ defined and analytic in $V_0 \setminus \{\zeta_0\}$. By the Cauchy integral formula, for each $\zeta \in \mathbb{D}_r(\zeta_0)$, we have

$$\phi_{m_k}(\zeta) = \frac{1}{2\pi i} \oint_\gamma \frac{\phi_{m_k}(z)\, dz}{z - \zeta} \to \frac{1}{2\pi i} \oint_\gamma \frac{\phi(z)\, dz}{z - \zeta}, \tag{7.10}$$

where the convergence is uniform on $\mathbb{D}_s(\zeta_0)$ with $s < r$. This concludes the proof of the claim.
Finally, if ϕ is a limit function of (ϕ_n), from the relations

$$\phi_n(f(z)) = \frac{f^{n+1}(z)}{f^n(\zeta_0)} \quad \text{and} \quad \phi_n(z) = \frac{f^n(z)}{f^n(\zeta_0,)},$$

we deduce

$$\frac{\phi_n(f(z))}{f^{n+1}(z)} = \frac{\phi_n(z)}{f^n(z)}.$$

This latter equality may be written in the form

$$\phi_n(f(z)) = \phi_n(z)\left(\frac{f(f^n(z)) - f(0)}{f^n(z) - 0}\right).$$

Thus considering a subsequence, if necessary, and passing to the limit, we obtain

$$\phi(f(z)) = \phi(z)f'(0). \tag{7.11}$$

As V_0 is invariant for f, it follows from the above equality that

$$\phi(f^{n+1}(z)) = \phi(f^n(z))f'(0),$$

for all $z \in V_0$. Passing to the limit as $n \to \infty$, we deduce that $\phi(0) = \phi(0)f'(0)$ and the proof concludes. □

7.4 Nonconstant limit functions

To finish the proof of the classification theorem, we need the following lemma.

Lemma 7.4.1. *Let $f : \mathbb{D} \to \mathbb{D}$ be a univalent function with the property that there exists a subsequence $(f^{n_k})_k$ such that $f^{n_k}(z) \to z$ locally uniformly on $\mathbb{D}$ as $k \to \infty$. Then f is surjective and, in consequence, f is a Möbius transformation.*

Proof. Choose $w \in \mathbb{D}$ and consider the sets $M_n = \{z : f^n(z) = w\}$. Since these sets are countable, so is $M = \bigcup_n M_n$. Take $r \in (|w|, 1)$ such that the circle γ of radius r centered at the origin does not pass through any point in M. Now define $n(\gamma, f^n - w)$ as the number of points in the set $M_n \cap \mathbb{D}_r$. As the f^n's are univalent, it follows from the analytic continuation principle that each $n(\gamma, f^n - w)$ is finite. By the argument principle, we have

$$n(\gamma, f^{n_k} - w) = \frac{1}{2\pi i}\oint_{\gamma} \frac{(f^{n_k})'(z)}{f^{n_k}(z) - w}\,dz.$$

Since $z = \lim_{k\to\infty} f^{n_k}(z)$ uniformly on γ, we deduce that $\lim_{k\to\infty} n(\gamma, f^{n_k} - w) = 1$ and, therefore, there exists a $k_0 \in \mathbb{N}$ such that $n(\gamma, f^{n_k} - w) = 1$ for all $k \geq k_0$. In particular, this shows that there is a $z \in \mathbb{D}_r$ such that $f^{k_0}(z) = w$, and the proof concludes. □

Theorem 7.4.2. *Let f be a transcendental entire function and let U be an invariant Fatou component such that $\mathbb{L}(U)$ contains some nonconstant function. Then the following statements hold:*

(1) *The identity map I belongs to $\mathbb{L}(U)$.*
(2) *Function f is univalent and surjective in U.*

(3) *Every nonconstant limit function is a conformal isomorphism from U onto itself.*
(4) *The set $\mathbb{L}(U)$ does not contain any constant limit function.*

Proof. (1) Let ϕ be a nonconstant limit function. Then there exists a subsequence $(f^{n_j})_j$ convergent to ϕ uniformly on compact subsets of U. To prove that $\phi(U) \subset U$, fix $w \in U$ and consider the function $\phi(z) - \phi(w)$ that is analytic and not identically null. Thus the zeros are isolated and, therefore, there exists a disc $\mathbb{D}(w)$ with center w such that $\overline{\mathbb{D}}(w) \subset U$ and $\phi(z) \neq \phi(w)$ for all $z \in \partial(\mathbb{D}(w))$. Hence

$$r = \min\{|\phi(z) - \phi(w)| : z \in \partial(\mathbb{D}(w))\} > 0.$$

Choose a $j_0 \in \mathbb{N}$ such that

$$|f^{n_j}(z) - \phi(z)| < r \quad \text{for } j \geq j_0 \text{ and } z \in \partial(\mathbb{D}(w)). \tag{7.12}$$

Applying Rouche's theorem to the functions $f^{n_j} - \phi$ and $\phi - \phi(w)$, we deduce that $\phi - \phi(w)$ and $f^{n_j} - \phi(w)$ have the same number of zeros in the disc $\mathbb{D}(w)$. Thus there is a $z_j \in \mathbb{D}(w)$ such that $f^{n_j}(z_j) = \phi(w)$ for all $j \geq j_0$. As $f(U) \subset U$, this implies that $\phi(w) \in U$. So we have proved that

$$\phi(U) \subset U. \tag{7.13}$$

By passing to a subsequence, if necessary, we may assume that

$$m_j = n_j - n_{j-1} \to \infty. \tag{7.14}$$

This $(f^{m_j})_j$ is normal in U and we may suppose that it converges uniformly on compact subsets of U to a function ψ. This function may be analytic in U or identically equal to ∞ and, in consequence, we must use the chordal metric to prove that $\psi(\phi(z)) = \phi(z)$ for all $z \in U$. Given $z \in U$, we have

$$d_c(\psi(\phi(z)), \phi(z)) \leq d_c(\psi(\phi(z)), f^{n_j}(z)) + d_c(f^{n_j}(z), \phi(z)).$$

This inequality may be written in another form

$$d_c(\psi(\phi(z)), \phi(z)) \leq d_c(\psi(\phi(z)), f^{m_j + n_{j-1}}(z)) + d_c(f^{n_j}(z), \phi(z)). \tag{7.15}$$

For a fixed $\epsilon > 0$, there exists a $j_0 \in \mathbb{N}$ such that $d_c(f^{n_j}(z), \phi(z)) < \epsilon/3$ for $j \geq j_0$. Therefore, (7.15) yields

$$d_c(\psi(\phi(z)), \phi(z)) \leq d_c(\psi(\phi(z)), \psi(f^{n_{j-1}}(z))) + d_c(\psi(f^{n_{j-1}}(z)), f^{m_j}(f^{n_{j-1}}(z))) + \frac{\epsilon}{3}. \tag{7.16}$$

By continuity, there exists a $\delta > 0$ such that

$$d_c(\phi(z), w) < \delta \quad \Longrightarrow \quad d_c(\psi(\phi(z)), \psi(w)) < \frac{\epsilon}{3}. \tag{7.17}$$

Choose $j_0^* \in \mathbb{N}$ such that $d_c(\phi(z), f^{n_j}(z)) < \delta$ for $j \geq j_0^*$. Then, for $j > j_0^*$,

$$d_c(\psi(\phi(z)), \psi(f^{n_{j-1}}(z))) < \frac{\epsilon}{3}. \tag{7.18}$$

By (7.13), $K = \{f^{n_j}(z) : j \in \mathbb{N}\} \cup \{\phi(z)\}$ is a compact subset of U; thus there exists a $j_0^{**} \in \mathbb{N}$ such that

$$d_c(\psi(w), f^{m_j}(w)) < \frac{\epsilon}{3} \quad \text{for } w \in K, j \geq j_0^{**}.$$

In particular, we have

$$d_c(\psi(f^{n_{j-1}}(z)), f^{m_j}(f^{n_{j-1}}(z))) < \frac{\epsilon}{3} \quad \text{for } j \geq j_0^{**}. \tag{7.19}$$

It follows from (7.16), (7.18), and (7.19) that $d_c(\psi(\phi(z)), \phi(z)) < \epsilon$ for all $\epsilon > 0$. This yields the equality

$$\psi(\phi(z)) = \phi(z) \quad \text{for all } z \in U. \tag{7.20}$$

The above relation implies that ψ cannot be the constant function ∞. Thus ψ is analytic in U. As $\psi(w) = w$ for $w \in \phi(U)$, by the identity principle, ψ must be the identity I in U, and this proves (1).

(2) Let $z, w \in U$ be such that $f(z) = f(w)$, then

$$f^{m_j}(z) = f^{m_j-1}(f(z)) = f^{m_j-1}(f(w)) = f^{m_j}(w).$$

Since $f^{m_j} \to I$ on U, we conclude that $I(z) = I(w)$. Thus f is injective in U. By Corollary 7.2.5, U is simply connected and we may combine Lemma 7.4.1 with Riemann's theorem to deduce that f is surjective.

(3) and (4) Let ϕ be a nonconstant limit function and suppose that $f^{n_j} \to \phi$ uniformly on compact subsets of U. We denote by $g : U \to U$ the inverse of f and note that (g^n) is normal in U (each g^n does not take any value $z \in \mathbb{J}(f)$). Then there is a subsequence of $(g^{n_k})_k$ which is uniformly convergent on compact subsets of U to a function ψ. To simplify, we suppose that the subsequence is the same $(g^{n_k})_k$. Given $z \in U$, we prove that

$$\psi(\phi(z)) = \lim_{k\to\infty} g^{n_k} \circ f^{n_k}(z). \tag{7.21}$$

By the continuity of ψ at the point $\phi(z)$, given $\epsilon > 0$, there is a $\delta > 0$ such that $d_c(\psi(\phi(z)), \psi(w)) < \epsilon$ whenever $d_c(\phi(z), w) < \delta$. Now choose a $k_0 \in \mathbb{N}$ such that $d_c(\phi(z), f^{n_k}(z)) < \delta$ for $k \geq k_0$. Therefore, we have

$$d_c(\psi(\phi(z)), \psi(f^{n_k}(z))) < \epsilon \quad \text{for } k \geq k_0. \tag{7.22}$$

As $\{f^{n_k}(z) : k \in \mathbb{N}\}$ is relatively compact, there exists a k_0^* such that $d_c(\psi(f^{n_h}(z)), g^{n_k}(f^{n_h}(z))) < \epsilon$ for $k \geq k_0^*$ and $h \in \mathbb{N}$. This, together with (7.22), yields

$$d_c(\psi(\phi(z)), g^{n_k}(f^{n_k}(z))) < 2\epsilon \quad \text{for } k \geq \max\{k_0, k_0^*\},$$

for every $\epsilon > 0$. Thus we have proved (7.21). Finally, having in mind that $g^{n_k} \circ f^{n_k}$ is the identity map in U, we conclude that $\psi(\phi(z)) = z$ for every $z \in U$. On the one hand, this implies that ψ and ϕ are not constant in U and, therefore, (4) is proved. On the other hand, the above relation also implies that ϕ must be injective in U. We can reverse the roles of ϕ and ψ and deduce that $\phi(\psi(z)) = z$ in U. Now (3) follows easily. □

Theorem 7.4.3. *Let f be a transcendental entire function and let U be a Fatou invariant component for f. If $\mathbb{L}(U)$ contains some nonconstant limit function, then U is a Siegel disc.*

Proof. By Theorem 7.4.2, $f : U \to U$ is bijective. On the other hand, Corollary 7.2.5 tells us that U is simply connected (different from $\mathbb{C}$ because $\mathbb{J}(f)$ is nonempty). If $\psi : U \to \mathbb{D}$ is a conformal isomorphism, the conjugate of f given by $F = \psi \circ f \circ \psi^{-1}$, is necessarily a Möbius transformation. We have seen in Chapter 1 that if a Möbius transformation is not conjugate to $S(z) = \lambda z$ where $|\lambda| = 1$, then all orbits are convergent to a point. In our case this is not possible because $\mathbb{L}(U)$ contains some nonconstant limit function. Therefore, F must be conjugate to a Möbius transformation of the form $S(z) = \lambda z$, where $\lambda = e^{2\pi\theta i}$. If θ is rational, then there is a $p \in \mathbb{N}$ such that $S^p(z) = z$ for all z. That is, f^p is the identity map which is not possible because f is transcendental. Hence θ must be irrational and, therefore, U is a Siegel disc. □

The proof of the next theorem is Exercise 4.

Theorem 7.4.4. *Let f be a transcendental entire function. If $F(f)$ has a Siegel cycle, then $\mathbb{F}(f)$ has preperiodic components.*

7.5 Baker domains

We have seen that transcendental entire functions have no Herman rings. Another important difference between the dynamics of rational functions and that of entire transcendental functions is that, in the latter, there may be Baker and wandering domains.

The following example is due to Fatou [41].

Example 7.5.1. Let $f(z) = 1 + z + e^{-z}$. The critical points of f are $z = 2n\pi i$, with $n \in \mathbb{Z}$. We will prove that the orbits starting in the right half-plane $\mathbb{R}e(z) > 0$ are divergent. If $\mathbb{R}e(z) > 0$, we have

$$\mathbb{R}e(f(z)) = 1 + \mathbb{R}e(z) + e^{-\mathbb{R}e(z)} \cos(\mathbb{I}m(z)) \geq \mathbb{R}e(z) + (1 - e^{-\mathbb{R}e(z)}),$$

where the term in parentheses is positive. This shows that this half-plane is invariant for f and we can apply the above relation to $f(z)$:

$$\mathbb{R}e(f^2(z)) \geq \mathbb{R}e(f(z)) + (1 - e^{-\mathbb{R}e(f(z))}).$$

In an obvious way, this yields

$$\mathbb{R}e(f^2(z)) \geq \mathbb{R}e(z) + 2(1 - e^{-\mathbb{R}e(z)}).$$

In general, we have

$$\mathbb{R}e(f^n(z)) \geq \mathbb{R}e(z) + n(1 - e^{-\mathbb{R}e(z)}),$$

which implies that $\lim_{n\to\infty} f^n(z) = \infty$ for $\mathbb{R}e(z) > 0$. Note that if U is the Fatou component containing $\mathbb{R}ez > 0$, then $f(U) \subset U$.

The name "Baker domain" was introduced by Eremenko and Lyubich [39]. Before that, names such as *domain of attraction of* ∞ and *essentially parabolic domain* were used.

There are different ways to obtain examples of Baker domains. A common way is the following:

Given an entire function g such that g is a self-function of $\mathbb{C}^*$, a *logarithmic lift* of g is an entire function f verifying

$$e^{f(z)} = g(e^z).$$

If we put $\pi(z) = e^z$, then the above equality is the same as $\pi(f(z)) = g(\pi(z))$. Note that $g(\pi(z)) \neq 0$ for all $z \in \mathbb{C}$, therefore, we may consider an analytic branch of $\log g(\pi(z))$ defined up a complex constant on the whole $\mathbb{C}$. Thus $f(z) = \log g(\pi(z))$ is a logarithmic lift of g. In applications, g can be expressed as $g(z) = z^n e^{\phi(z)}$, where ϕ is some entire function. The logarithmic lifts play an important role in the study of bounded singular-type functions.

To find examples of Baker domains, we start with an entire function g whose behavior near zero is known. If $\pi(z) = e^{az}$ with $a \neq 0$, we suppose that $g(\pi(z)) \neq 0$ for every $z \in \mathbb{C}$. Then, we lift the dynamics of g to $\mathbb{C}$ taking an entire function f such that $\pi \circ f = g \circ \pi$. The idea is to study the behavior of the orbits under f knowing the behavior of the orbits under g close to 0.

Example 7.5.2. Take $g(w) = cwe^{-w}$ with $c \neq 0$ and $\pi(z) = e^{-z}$. Then $g(\pi(z)) \neq 0$ for all $z \in \mathbb{C}$ and the function $f(z) = -\log c + z + e^{-z}$ satisfies the equality $\pi(f(z)) = g(\pi(z))$ for all $z \in \mathbb{C}$. If we assume that $|c| < 1$, then the origin is an attracting fixed point of g and there is an open neighborhood V of 0 such that $g^n(w) \to 0$ for every $w \in V$ (take V small enough so that $g(V) \subset V$). Therefore, $\pi \circ f^n(z) \to 0$ for every $z \in \pi^{-1}(V)$ and, in consequence, $f^n(z) \to \infty$. Hence, the Fatou component of f containing the open set $\pi^{-1}(V)$ is a Baker domain (notice that $\pi^{-1}(V)$ is invariant under f).

7.6 Wandering domains

We need the next proposition to obtain an example of wandering domain.

Proposition 7.6.1. *Let g be an entire function such that $g(z+c) = g(z) + c$ for all $z \in \mathbb{C}$, with c being a complex constant. The following statements hold:*
(i) *If $h(z) = z + c$, then*

$$h(\mathbb{F}(g)) = h^{-1}(\mathbb{F}(g)) = \mathbb{F}(g).$$

(ii) *If $f(z) = g(z) + c$ for all $z \in \mathbb{C}$, then $\mathbb{J}(f) = \mathbb{J}(g)$.*

Proof. The following relations are evident:

$$(1) \quad g^n(z+c) = g^n(z) + c \quad \text{and} \quad (2) \quad g(z-c) = g(z) - c.$$

(i) If $z_0 \in \mathbb{F}(g)$, we choose a neighborhood $U \subset \mathbb{F}(g)$ of z_0 where (g^n) is normal. By (1), we have $g^n(h(z)) = g^n(z) + c$. Thus (g^n) is normal in $h(U)$ and, in consequence, $h(z_0) \in \mathbb{F}(g)$. Now using (2) and by the same argument, we deduce that $h^{-1}(z_0) \in \mathbb{F}(g)$. On the other hand, it is clear that

$$\mathbb{F}(g) = h \circ h^{-1}(\mathbb{F}(g)) \subset h(\mathbb{F}(g)).$$

(ii) Let z_0 be a repelling p-periodic point for g. We shall prove that $z_0 \in \mathbb{J}(f)$. By Marty's theorem, to show that (f^n) is not normal in a neighborhood of z_0, we must prove that the quotient

$$\frac{|(f^{pn})'(z_0)|}{1 + |f^{pn}(z_0)|^2} \tag{7.23}$$

tends to ∞ as $n \to \infty$. First, we prove by induction the equality $f^n(z) = g^n(z) + nc$. Indeed, we have

$$f^{n+1}(z) = f(f^n(z)) = f(g^n(z) + nc) = g(g^n(z) + nc) + c = g^{n+1}(z) + (n+1)c.$$

Finally, note that

$$f^{pn}(z_0) = g^{pn}(z_0) + npc = z_0 + cnp \quad \text{and} \quad (f^{pn})'(z_0) = (g^{pn})'(z_0) = \lambda^n,$$

where $\lambda = (g^p)'(z_0)$ is the multiplier of the repelling point z_0 and, therefore, $|\lambda| > 1$. Thus the quotient (7.23) adopts the form

$$\frac{|(f^{pn})'(z_0)|}{1 + |f^{pn}(z_0)|^2} = \frac{|\lambda|^n}{1 + |z_0 + cpn|^2}.$$

Obviously, the above quotient tends to ∞ as $n \to \infty$ and we conclude that $\mathbb{J}(g) \subset \mathbb{J}(f)$. The reverse inclusion follows by the same argument swapping the roles of f and g. □

Example 7.6.2. Let $g(z) = z+1-e^z$. The critical points of g are $z = 2\pi ni$, with $n \in \mathbb{Z}$. These points are also fixed points for g (superattracting). Consider the function $f(z) = g(z)+2\pi i$. Then g and f satisfy the conditions of Proposition 7.6.1 and, therefore, $\mathbb{F}(g) = \mathbb{F}(f)$. If we denote by $U_{2\pi ni}$ the component of $\mathbb{F}(g)$ that contains the point $z_n = 2\pi ni$, then each $U_{2\pi ni}$ is a wandering domain for f because $f(2\pi ni) = 2(n+1)\pi i$.

7.7 Exercises

1. (Baker) Let (a_n) be a sequence of positive real numbers such that $2a_n < a_{n+1}$ and consider the function

$$h(z) = \prod_{n=1}^{\infty}\left(1 - \frac{z}{a_n}\right).$$

   If $f(z) = z + 2\pi i + h(e^z)^2$, prove that $\mathbb{F}(f)$ has infinitely many wandering domains.
2. Let $f(z) = z + \log c + e^z$, where $0 < |c| < 1$. Show that f has a Baker domain V containing a left half-plane.
3. Prove that the function $f(z) = 2z + e^z$ has a Baker domain that contains a half-plane $H = \{z : \mathbb{R}e(z) < R\}$.
4. Let f be a transcendental entire function. If $F(f)$ has a Siegel cycle, then $\mathbb{F}(f)$ has preperiodic components.
5. Let $f(z) = z + (\lambda - 1)(e^z - 1) + 2\pi i$, where $\lambda = e^{2\pi\theta i}$, with θ a Diophantine number. Show that $\mathbb{F}(f)$ has a simply connected wandering domain U such that f is univalent on U.
6. Without applying the classification theorem, prove that an entire function has no Herman rings.
7. Prove this simple criterion for a given Fatou component to be non-wandering in the framework of rational maps: "Suppose that V is a Fatou component of a rational function f and $f^{n_k} \to \phi$ locally uniformly in V. Then $\phi(V) \cap \mathbb{F}(f) \neq \emptyset$ implies that V is preperiodic. Hint: use the fact that, if f is rational, then the sequence $(f^n(W))$ either consists of mutually disjoint sets or there exist m and n such that $f^{n+m}(W) = f^m(W)$ for any Fatou component W".
8. Let $P(z)$ be a polynomial of degree $n \geq 2$ with the property that the Julia set $\mathbb{J}(P)$ contains a Jordan curve γ. Show that the bounded region limited by γ is a Fatou component.
9. Let P be a polynomial of degree $n \geq 2$ and z^* an attracting fixed point of P. Suppose that $\mathbb{F}(P)$ has a component $U \neq \Omega_0(P, z^*)$ which is mapping onto $\Omega_0(P, z^*)$ by P. Prove that there must be infinitely many components in $\mathbb{F}(P)$.

Bibliography remarks

In Section 7.2, we follow [8].

Sections 7.3–7.4 are inspired by [47, Section 4.4]. In this book, the proof of Theorem 7.3.4 is rather incomplete and, without our Lemma 7.4.1, the proof of the classification theorem is not complete.

Section 7.5 is based on [78] and [63, Section 3.4].

Section 7.6 is inspired by [63, Section 3.4].

8 Singular values

In this chapter, we provide evidence that forward orbits of singular values of an entire function determine the general features of the global dynamics of the function. Among other results, we show that attracting or rationally indifferent cycles attract the forward orbit of some singular value of the function. This allows us to obtain quantitative information about the dynamics of the function, especially in the case the set of singular values is finite or bounded.

8.1 Singular values

In the framework of rational functions, there is a relevant relationship between critical points and attracting periodic points: if f is a rational map of degree $d \geq 2$, then in the attractive basin of every attracting cycle there is a critical point of f. This is not the case for transcendental entire functions. We may consider, for example, the function $f(z) = \lambda e^z$ for $\lambda \in (0, 1/e)$. In the next chapter, we will see that f has an attracting fixed point and, nevertheless, f has no critical point. We will show that, for transcendental entire functions, the set of "critical" points must be extended.

Definition 8.1.1. Let $w_0 \in \mathbb{C}$. We will say that w_0 is a nonsingular value of f if there is a neighborhood V of w_0 such that every branch of f^{-1} in V is well defined and analytic.

According to [66] singular values of f (also called singularities of f^{-1}) are of the following types:

(1) Algebraic singularities. They are the critical values of f. Recall that w_0 is said to be a *critical value* of f if there is z_0 such that $f'(z_0) = 0$ and $w_0 = f(z_0)$. We denote the set of all critical values of f by $CV(f)$.

(2) Transcendental singularities. They are the asymptotic values of f. We say that w_0 is an *asymptotic value* of f if there is a curve $\gamma(t)$ satisfying

$$\lim_{t\to+\infty} \gamma(t) = \infty \quad \text{and} \quad \lim_{t\to+\infty} f(\gamma(t)) = w_0.$$

The set of all asymptotic values of f is denoted by $AV(f)$. It is known that a Picard exceptional value of f is an asymptotic value of f (see [46, 50]).

(3) Limit points of type (1) and (2).

Then $\overline{AV(f) \cup CV(f)}$ is the set of singular values of f and it will be denoted by $\operatorname{sing}(f^{-1})$. Notice that if f is an entire function and the region $\Omega \subset \mathbb{C}$ contains no critical or asymptotic values then $f^{-1} : f^{-1}(\Omega) \to \Omega$ is a covering map (that is, an open surjective map that is locally an analytic isomorphism).

Examples 8.1.2. (1) If f is rational, the only singular values are the critical values of f. Therefore, rational functions have finitely many singular values.

https://doi.org/10.1515/9783111689685-008

(2) In the case of the exponential function $E_\lambda(z) = \lambda e^z$, for every curve $\gamma(t)$ such that $\mathbb{R}e(\gamma(t)) \to -\infty$, we have

$$\lim_{t\to+\infty} |E_\lambda(\gamma(t))| = \lambda \lim_{t\to+\infty} e^{\mathbb{R}e(\gamma(t))} = 0.$$

Then 0 is an asymptotic value for E_λ and it is easy to show that every $a \neq 0$ is nonsingular.

If we consider E_λ as a function from $\mathbb{C}$ into $\hat{\mathbb{C}}$, then ∞ also is an asymptotic value. Indeed, if $\gamma(t)$ is a curve such that $\mathbb{R}e(\gamma(t)) \to +\infty$, then $E_\lambda(\gamma(t)) \to \infty$.

(3) If $f(z) = ze^z$, then $z = -1$ is the unique critical point of f. Let us see that every nonnull a is not an asymptotic value of f. Suppose that there is a curve $\gamma(t) = x(t) + iy(t) \to \infty$ such that $f(\gamma(t)) \to a \neq 0$. This implies that $|\gamma(t)|e^{x(t)} \to |a|$. Therefore, we deduce that $e^{x(t)} \to 0$ and, in consequence, $x(t) \to -\infty$. On the other hand, if $a = a_1 + ia_2$, we have

$$\begin{cases} x(t)e^{x(t)}\cos(y(t)) - y(t)e^{x(t)}\sin(y(t)) \to a_1, \\ x(t)e^{x(t)}\sin(y(t)) + y(t)e^{x(t)}\cos(y(t)) \to a_2. \end{cases} \tag{8.1}$$

As $x(t)e^{x(t)} \to 0$, it follows that

$$y(t)e^{x(t)}\sin(y(t)) \to -a_1 \quad \text{and} \quad y(t)e^{x(t)}\cos(y(t)) \to a_2$$

and then

$$y(t)^2e^{2x(t)}\sin^2(y(t)) \to a_1^2 \quad \text{and} \quad y(t)^2e^{2x(t)}\cos^2(y(t)) \to a_2^2. \tag{8.2}$$

This yields $y(t)^2e^{2x(t)} \to |a|^2$. If we put $\phi(t) = y(t)^2e^{2x(t)}$, then $y(t)^2 = \phi(t)e^{-2x(t)}$, with $\phi(t) \to |a|^2$. Note that $y(t) \to \infty$. Finally, substituting this expression of $y(t)^2$ into (8.2), we obtain

$$\phi(t)\sin^2(y(t)) \to a_1^2 \quad \text{and} \quad \phi(t)\cos^2(y(t)) \to a_2^2.$$

Obviously, we deduce

$$\sin^2(y(t)) \to \frac{a_1^2}{|a|^2} \quad \text{and} \quad \cos^2(y(t)) \to \frac{a_2^2}{|a|^2}.$$

Now an elementary argument, using the intermediate value theorem and the fact that $y(t) \to \infty$, allows us to prove that both limits, $\lim_{t\to\infty}\sin^2(y(t))$ and $\lim_{t\to\infty}\cos^2(y(t))$, do not exist. Therefore, only the origin may be an asymptotic value.

Finally, in view of (8.1), $f \circ \gamma(t) \to 0$ for every curve $\gamma(t) = x(t) + iy(t)$ such that $x(t) \to -\infty$ and $y(t)$ is bounded.

(4) Let g be an entire function such that $g(z) \neq 0$ for every $z \in \mathbb{C}^*$. If f is a logarithmic lift of g, then we have

$$z \in \operatorname{sing}(f^{-1}) \quad \Longrightarrow \quad e^z \in \operatorname{sing}(g^{-1}).$$

Next, we show the relationship between $\operatorname{sing}(f^{-1})$ and the singular sets of the iterates f^n.

Theorem 8.1.3. $\overline{\bigcup_{n\geq 1} f^n(\operatorname{sing}(f^{-1}))} = \overline{\bigcup_{n\geq 1} \operatorname{sing}((f^n)^{-1})}$.

Proof. (1) $\bigcup_{n\geq 1} f^n(\operatorname{sing}(f^{-1})) \subset \bigcup_{n\geq 1} \operatorname{sing}((f^n)^{-1})$. If $a = f^n(b)$ with $b \in \operatorname{sing}(f^{-1})$, the following possibilities may occur:

(1a) b is a critical value of f. Then there exists a c such that $b = f(c)$ and $f'(c) = 0$. Thus $a = f^n(b) = f^{n+1}(c)$ and

$$(f^{n+1})'(c) = (f^n)'(f(c))f'(c) = 0.$$

We conclude that a is a critical value of f^{n+1}.

(1b) b is an asymptotic value of f. Then there is a $\gamma(t) \to \infty$ such that $f(\gamma(t)) \to b$. Thus

$$f^{n+1}(\gamma(t)) = f^n(f(\gamma(t)) \to f^n(b) = a,$$

which tells us that a is an asymptotic value of f^{n+1}.

(1c) $b = \lim_{k\to\infty} b_k$, where each b_k is a critical or asymptotic value of f. Applying (1a) or (1b) to $a_k = f^n(b_k)$, we deduce that a_k is a critical or asymptotic value of f^{n+1}. As $\lim_{k\to\infty} a_k = \lim_{k\to\infty} f^n(b_k) = f^n(b) = a$, it follows that a belongs to $\operatorname{sing}((f^{n+1})^{-1})$.

(2) $\bigcup_{n\geq 1} \operatorname{sing}((f^n)^{-1}) \subset \overline{\bigcup_{n\geq 1} f^n(\operatorname{sing}(f^{-1}))}$. Let $a \in \operatorname{sing}((f^n)^{-1})$. Again we distinguish three cases:

(2a) a is a critical value of f^n. There exists a b such that $f^n(b) = a$ and $(f^n)'(b) = 0$. Notice that

$$(f^n)'(b) = \prod_{j=0}^{n-1} f'(f^j(b)) = 0.$$

Then there is a $j \leq n-1$ such that $f'(f^j(b)) = 0$. Therefore, $f^{j+1}(b)$ is a critical value of f and $a = f^{n-j-1}(f^{j+1}(b))$. This implies that a is in $f^{n-j-1}(\operatorname{sing}(f^{-1}))$.

(2b) a is an asymptotic value of f^n. Then, there is a $\gamma(t) \to \infty$ such that $f^n(\gamma(t)) \to a$. If we denote by j $(0 \leq j < n)$ the largest integer such that $f^j \circ \gamma$ is unbounded, then $f^{j+1} \circ \gamma$ is bounded. We will show that $f^j \circ \gamma \to \infty$ and $f^{j+1} \circ \gamma \to c$, where c is a solution of the equation $f^{n-j-1}(z) = a$. By hypothesis, given $\varepsilon > 0$, there exists a $T > 0$ such that $|f^n(\gamma(t)) - a| < \varepsilon$ for $t > T$. In particular, we have

$$\left|f^{n-j}(f^j \circ \gamma(t)) - a\right| < \varepsilon \quad \text{for } t > T. \tag{8.3}$$

For a fixed $R > 0$, the equation $f^{n-j}(z) = a$ has only a finite number of solutions in $\overline{\mathbb{D}}_R$, say $z_1, \ldots, z_p$. We may choose an $s > 0$ sufficiently small such that the discs $\mathbb{D}_s(z_j)$ are mutually disjoint. By the maximum modulus principle applied to $1/(f^{n-j}(z) - a)$ in $\Omega = \overline{\mathbb{D}}_R \setminus (\bigcup_j \mathbb{D}_s(z_j))$, there is an $m > 0$ such that

$$|f^{n-j}(z) - a| > m \quad \text{for } z \in \Omega. \tag{8.4}$$

Taking $\varepsilon = m$, (8.3) and (8.4) imply that $f^j(\gamma(t)) \notin \Omega$ for $t > T$. Note that, since $f^j \circ \gamma_{t>T}$ cannot visit any of the discs $\mathbb{D}_s(z_j)$, $f^j(\gamma(t))$ is in the exterior of the disc $\overline{\mathbb{D}}_R$ for $t > T$. This shows that $\gamma_j = f^j \circ \gamma$ leaves any compact subset of $\mathbb{C}$ and then converges to infinity. Now notice that

$$f^{n-j-1} \circ (f \circ \gamma_j)(t)) \to a.$$

As $f \circ \gamma_j$ is bounded and, therefore, contained in some compact, by a similar argument, it is easy to prove that $f \circ \gamma_j(t) \to c$, where $c \in \mathbb{C}$ so that $f^{n-j-1}(c) = a$ (this is Exercise 13). Then we conclude that $a \in f^{n-j-1}(\text{sing}(f^{-1}))$.

(2c) $a = \lim_{k\to\infty} a_k$, where $a_k \in CV(f^n) \cup AV(f^n)$ for all k. According to (2a) and (2c), for each k there is a $j(k) \leq n-1$ such that $a_k \in f^{n-j(k)-1}(\text{sing}(f^{-1}))$. Since $(j(k))_k$ is an infinite sequence with a finite number of distinct entries, there is a subsequence $(j(k_n))_n$ such that $j(k_n) = j$ for every $n \in \mathbb{N}$ and $j < n$. Obviously, $a_{k_n} \to a$ and, in consequence, $a \in \overline{f^{n-j-1}(\text{sing}(f^{-1}))}$. □

For every entire function f, we denote by $S^+(f)$ the set defined by

$$S^+(f) = \overline{\bigcup_n f^n(\text{sing}(f^{-1}))}.$$

Then $S^+(f)$ is the closure of all forward orbits of singular values and is called the *postsingular set* of f.

The laborious proof of the next Theorem can be consulted in [55, Section 13.3].

Theorem 8.1.4. *Let f be a transcendental entire function. If V is an open simply connected set disjoint from $S^+(f)$, then for every $w \in V$, $n \in \mathbb{N}$ and $z \in f^{-n}(w)$ there is a unique analytic branch of f^{-n} defined on V so that $f^{-n}(w) = z$.*

Theorem 8.1.5 (Expansivity of $\mathbb{J}(f)$). *Let f be a transcendental entire function and let $K \subset \mathbb{C}$ be a compact set such that $K \cap \text{sing}(f^{-1}) = \emptyset$. If $z_0 \in \mathbb{J}(f)$ and $D \subset \mathbb{C}$ is an open connected neighborhood of z_0, then there exists an $N \in \mathbb{N}$ such that $K \subset f^n(D)$ for all $n \geq N$.*

Proof. Arguing by contradiction, suppose that there exists a strictly increasing sequence of positive integers $(n_j)_j$ such that $K \setminus f^{n_j}(D) \neq \emptyset$ for each $j \in \mathbb{N}$. Choose $a_j \in K \setminus f^{n_j}(D)$. Since K is compact, we can suppose that $a_j \to a \in K$. As a is nonsingular (recall that Picard exceptional values are singular), there exist $a_1, a_2 \in f^{-1}(a)$ with $a_1 \neq a_2$ and two branches of f^{-1}, f_1^{-1} and f_2^{-1}, defined and analytic in a neighborhood U of a satisfying

$$f_1^{-1}(U) \cap f_2^{-1}(U) = \emptyset \quad \text{and} \quad f_1^{-1}(a) = a_1,\ f_2^{-1}(a) = a_2.$$

To simplify, we assume that $a_j \in U$ for all $j \in \mathbb{N}$. If we put $a_j^1 = f_1^{-1}(a_j)$ and $a_j^2 = f_2^{-1}(a_j)$, then $a_j^1 \to a_1$ and $a_j^2 \to a_2$.

For each j, we know that $f^{n_j}(z) \neq a_j$ for all $z \in D$. Therefore

$$f^{n_j-1}(z) \neq a_j^1, a_j^2 \quad \text{for } z \in D.$$

There exists a $j_0 \in \mathbb{N}$ such that

$$|a_j^1| < 1 + |a_1|, \quad |a_j^2| < 1 + |a_2|, \quad \text{and} \quad |a_j^1 - a_j^2| > (1/2)|a_1 - a_2|,$$

for $j \geq j_0$. Thus, applying Theorem 2.2.4 to (f^{n_j-1}), we deduce that (f^{n_j-1}) is normal in D. Finally, it follows from Theorem 6.3.9 that (f^n) is normal in D, which is a contradiction. □

Given a region $\Omega \subset \mathbb{C}$, as usual, we denote by $H(\Omega)$ the set of analytic functions defined on Ω. If f is an entire function, consider the family $\mathcal{F}$ defined by

$$\mathcal{F} = \{g \in H(\Omega) : \exists m \text{ such that } g \text{ is an analytic branch of } (f^m)^{-1}\}.$$

Lemma 8.1.6. *Let f be a transcendental entire function and let $\Omega \subset \mathbb{C}$ be a region. If $\mathcal{F}$ is the above family, consisting of all analytic branches of some $(f^n)^{-1}$ on Ω, then $\mathcal{F}$ is normal on Ω.*

Proof. According to Theorem 4.4.3, f has infinitely many periodic points of period n, except for at most one integer n. Then we can choose periodic points u and v with periods p and q, respectively, satisfying $p, q \geq 2$ and $O_f^+(u) \cap O_f^+(v) = \emptyset$. We prove that $\mathcal{F}$ is normal in $\Omega \setminus O_f^+(u)$ and $\Omega \setminus O_f^+(v)$, and then normal in Ω. Let us see that $\mathcal{F}$ is normal in $\Omega \setminus O_f^+(u)$. It suffices to show that every function belonging to $\mathcal{F}$ does not attain any value $w \in O_f^+(u)$. Indeed, suppose that $w \in O_f^+(u)$, $g \in \mathcal{F}$, and there is a $z \in \Omega \setminus O_f^+(u)$ such that $g(z) = w$. Let $m \in \mathbb{N}$ be such that $g = (f^m)^{-1}$. Then it is obvious that $f^m(w) = z$ and this implies that $z \in O_f^+(u)$, which is a contradiction. □

Lemma 8.1.7. *Let f be a transcendental entire function and let Ω be a relatively compact region that does not contain any singular value of f and such that $\Omega \cap \mathbb{J}(f) \neq \emptyset$. For each $n \in \mathbb{N}$, let h_n be an analytic branch of $(f^n)^{-1}$ defined on Ω. Then every limit of any subsequence of (h_n) is a constant.*

Proof. If not, there exists a subsequence of (h_j) uniformly convergent on compact subsets of Ω to a nonconstant function h. To simplify, we will denote the subsequence by (h_j). For each j, h_j is an analytic branch of some f^{n_j}. Choose $z_0 \in \Omega \cap \mathbb{J}(f)$ such that $h'(z_0) \neq 0$ and note that $h(z_0) \in \mathbb{J}(f)$ because $f^{n_j} \circ h_j(z_0) = z_0$ for all j and $\mathbb{J}(f)$ is closed and completely invariant. As the h_j's are univalent, Hurwitz's theorem asserts that h is also univalent. Take a disc $\mathbb{D}_r(z_0)$ contained in Ω and apply the 1/4-theorem of Koebe in the disc $\mathbb{D}_r(z_0)$ to the functions h_j's and h. Then we have

$$\mathbb{D}(h(z_0), r|h'(z_0)|/4) \subset h(\Omega) \quad \text{and} \quad \mathbb{D}(h_j(z_0), r|h_j'(z_0)|/4) \subset h_j(\Omega), \tag{8.5}$$

for all j. Now take a j_0 such that $|h_j'(z_0)| \geq (1/2)|h'(z_0)|$ for all $j \geq j_0$ and set $\delta = (1/16)r|h'(z_0)|$. There is a $j_1 \geq j_0$ such that $|h_j(z_0) - h(z_0)| < \delta$ for $j \geq j_1$. Then we may obtain an upper bound for $|h_j(z_0) - w|$:

$$|h_j(z_0) - w| \leq |h_j(z_0) - h(z_0)| + |h(z_0) - w| < 2\delta = (1/8)r|h'(z_0)| \leq (1/4)r|h_j'(z_0)|,$$

for $j \geq j_1$ and $w \in V = \mathbb{D}_\delta(h(z_0))$. It follows from (8.5) that $V \subset h_j(\Omega)$ for all $j \geq j_1$. Thus $f^{n_j}(V) \subset f^{n_j} \circ h_j(\Omega) = \Omega$ for all $j \geq j_1$, and this contradicts Theorem 8.1.5. □

The next result about the location of constant limits is due to Baker [6].

Theorem 8.1.8. *Let f be a transcendental entire function and $X = S^+(f) \cup \{\infty\}$. Every constant limit of a sequence $(f^{n_k})_k$ on a component of $\mathbb{F}(f)$ belongs to X.*

8.2 Singular values and Fatou components

In this section, among other results, we will see that the basin of attraction of an attracting or parabolic p-cycle always contains a singular value. We need the analytic continuation of an element $(f, \mathbb{D}_r(a))$ to the Mittag-Leffler star $\mathbb{S}_{ML}$ (here f is defined and analytic in the disc $\mathbb{D}_r(a)$). Let us explain briefly the basic facts (see, for instance, [60]).

Given an analytic element $(f, \mathbb{D}_r(a))$, the Mittag-Leffler star $\mathbb{S}_{ML}$ consists of those points of the complex plane which can be reached by an analytic continuation of $f(z)$ along all possible rays from the center a. If $z = a + re^{\theta i}$ $(0 \leq r < +\infty)$ is a ray on which there are points that cannot be reached this way, then there is a point $z_1 \neq a$ on the ray so that the element can be continued to any point of $[a, z_1)$ but not beyond. If a continuation is possible to any point of the ray, one puts $z_1 = \infty$. It is known that this star-like set $\mathbb{S}_{ML}$ is a simply connected region. Thus an analytic continuation in $\mathbb{S}_{ML}$ results in an analytic function $F(z)$ defined on $\mathbb{S}_{ML}$ and satisfying $F(z) = f(z)$ for every $z \in \mathbb{D}_r(a)$.

Our proof of the next theorem follows the ideas of Milnor's book (see [61, pp. 76–78]).

Theorem 8.2.1. *If $f(z)$ is a nonlinear entire function, then in the attractive basin of every attracting cycle there is some singular value of $f(z)$.*

Proof. By Theorem 8.1.3, it suffices to consider the case of a fixed point. Then we assume that $z(0)$ is an attracting fixed point for $f(z)$ and we will prove that there is a singular value of f in the immediate attractive basin of $z(0)$ (a critical or asymptotic value since the mentioned basin is open). We may assume that $z(0)$ is not superattracting because in this case $z(0)$ would be a critical value of f. So, according to Koenig's theorem, there exist a $\delta > 0$, a connected neighborhood W of $z(0)$, and a conformal isomorphism $\phi : W \to \mathbb{D}(\delta)$ satisfying

$$\phi(z(0)) = 0 \quad \text{and} \quad \phi(f(z)) = \lambda\phi(z) \quad \text{for all } z \in W,$$

with λ being the multiplier of $z(0)$. Remark 5.1.4 tells us that ϕ may be extended to the attractive basin $\Omega(f, z(0))$ such that $\phi(\Omega(f, z(0))) = \mathbb{D}(\delta)$, where the above equality is valid for $z \in \Omega(f, z(0))$. We denote by ψ the inverse of the restriction of ϕ to W. Now we proceed to continue analytically the element $(\psi, \mathbb{D}(\delta))$ in all directions from the origin. In this way, we obtain an analytic function defined from $\mathbb{S}_{ML}$ into $\mathbb{C}$, where $\mathbb{S}_{ML}$ denotes the Mittag-Leffler star of the element $(\psi, \mathbb{D}(\delta))$. First, we prove that $\psi(\mathbb{S}_{ML}) \subset \Omega_0(f, z(0))$. Indeed, $\psi(\mathbb{S}_{ML})$ is a region that contains W and, therefore, we just need to show that $\psi(\mathbb{S}_{ML}) \subset \Omega(f, z(0))$. The equality $f(\psi(w)) = \psi(\lambda w)$ holds for all $w \in \mathbb{D}(\delta)$. By the identity principle, it is valid for all $w \in \mathbb{S}_{ML}$. Then, given $w \in \mathbb{S}_{ML}$, we have $f^n(\psi(w)) = \psi(\lambda^n w)$ for all $n \in \mathbb{N}$ and, in consequence, we deduce that $f^n(\psi(w)) \to z(0)$ as $n \to \infty$. That is, $\psi(w)$ belongs to the attractive basin of $z(0)$. Now it is easy to show that $\mathbb{S}_{ML} \neq \mathbb{C}$. In fact, if $\mathbb{S}_{ML} = \mathbb{C}$, then ψ is a nonconstant entire function such that $\psi(\mathbb{C}) \subset \Omega_0(f, z(0))$ (recall that $\phi'(z(0)) = 1$). As $\mathbb{J}(f)$ is infinite, this contradicts the little theorem of Picard.

Thus there must exist some largest radius r such that ψ extends analytically throughout the disc $\mathbb{D}(r)$. Put $U = \psi(\mathbb{D}(r))$. Again U is a region containing W, so $U \subset \Omega_0(f, z(0))$. This ϕ is an isomorphism from U onto $\mathbb{D}(r)$ because (by the identity principle) the equalities $\phi \circ \psi(w) = w$ and $\psi \circ \phi(z) = z$ are valid in $\mathbb{D}(r)$ and U, respectively.

Claim. $\overline{U} \subset \Omega_0(f, z(0))$ and $\partial(U)$ contains a singular value of f.

If $z_1 \in \partial U$, for $z \in U$ as close to z_1 as one wishes, we have $\phi(f(z)) = \lambda\phi(z) \in \mathbb{D}(|\lambda| r)$. By continuity, it follows that $\phi(f(z_1)) \in \overline{\mathbb{D}}(|\lambda| r) \subset \mathbb{D}(r)$ and this yields $f(z_1) \in U \subset \Omega(f, z(0))$. As the basin is completely invariant, we conclude that $z_1 \in \Omega(f, z(0))$. Then $\overline{U} \subset \Omega(f, z(0))$ is a connected set containing $z(0)$ and this proves that $\overline{U} \subset \Omega_0(f, z(0))$.

Now, by contradiction, we prove that $\partial(U)$ contains a singular value of f (see Figure 8.1).

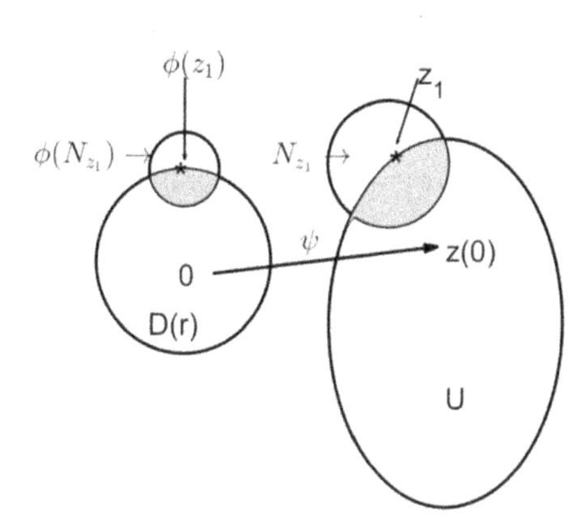

Figure 8.1: The isomorphism ψ from $D(r)$ onto U mapping 0 to $z(0)$.

So, we suppose that every point in $\partial(U)$ is nonsingular. Given $z_1 \in \partial U$, there exist neighborhoods $N^1_{z_1}$ and $N_{f(z_1)}$ of z_1 and $f(z_1)$, respectively, such that $f : N^1_{z_1} \to N_{f(z_1)}$ is a conformal isomorphism (both neighborhoods are contained in $\Omega(f, z(0))$). Let $N_{\phi(z_1)}$ be a neighborhood of $\phi(z_1)$ sufficiently small such that $\lambda N_{\phi(z_1)} \subset \mathbb{D}(r)$ ($|\lambda| < 1$).

By continuity, there exists a neighborhood $N_{z_1}^2$ of z_1 such that $\phi(N_{z_1}^2) \subset N_{\phi(z_1)}$. If we put $N_{z_1} = N_{z_1}^1 \cap N_{z_1}^2$, then $\phi(N_{z_1})$ is a neighborhood of $\phi(z_1)$ such that $\lambda\phi(N_{z_1}) \subset \mathbb{D}(r)$. Now we show that ψ may be extended by analytic continuation through the neighborhood $\phi(N_{z_1})$ composing the functions $L(z) = \lambda z$, ψ and f^{-1}. In fact, let $z \in \phi(N_{z_1})$ and $w = \psi(\lambda z)$. We have to prove that $w \in N_{f(z_1)}$, where f^{-1} is defined. Since $z = \phi(\eta)$ with $\eta \in N_{z_1}$, we have

$$w = \psi(\lambda z) = \psi(\lambda\phi(\eta)) = \psi \circ \phi \circ f(\eta) = f(\eta) \in f(N_{z_1}) \subset N_{f(z_1)}.$$

Thus the function $\hat{\psi}(z) = f^{-1} \circ \psi(\lambda z)$ is defined and analytic in $\phi(N_{z_1})$. Let us see that $\psi(z) = \hat{\psi}(z)$ for $z \in \phi(N_{z_1}) \cap \mathbb{D}(r)$: if $z \in \mathbb{D}(r) \cap \phi(N_{z_1})$, there exists $\zeta \in N_{z_1} \subset \Omega(f, z(0))$ such that $z = \phi(\zeta)$. As f^{-1} is a conformal isomorphism from $N_{f(z_1)}$ onto N_{z_1}, there is $\eta \in N_{f(z_1)}$ such that $\zeta = f^{-1}(\eta)$. Then $z = \phi(f^{-1}(\eta))$ and $\lambda z = \lambda\phi(f^{-1}(\eta)) = \phi(f(f^{-1}(\eta))) = \phi(\eta)$. This yields $\phi^{-1}(\lambda z) = \eta$ and it follows that $\hat{\psi}(z) = f^{-1}(\phi^{-1}(\lambda z)) = f^{-1}(\eta) = \zeta = \psi(z)$. Finally, if there were no singular values at all in ∂U, these local extensions would piece together to yield an analytic extension of ψ trough a disc which is strictly larger than $\mathbb{D}(r)$. □

There is an analogous result for parabolic cycles.

Theorem 8.2.2. *In the attractive basin of every parabolic cycle, there is some singular value.*

Proof. Again it suffices to consider the attractive basin of a parabolic fixed point z^* with $f'(z^*) = 1$. For a fixed petal P in the attractive basin, the key is to work with the Fatou function α defined in P and satisfying the Abel equation $\alpha(f(z)) = \alpha(z) + 1$. In this case, α^{-1} is defined in a right half-plane $\{w : \mathbb{R}e(w) > c\}$ and we try to extend α^{-1} leftwards.

Nevertheless, recalling Theorem 7.1.2, there is an alternative proof: If P contains no singular values of f, Theorem 7.1.2 tells us that $f(P) = P$ which contradicts Remarks 5.3.7(4). □

Theorem 8.2.3. *Let f be a transcendental entire function and*

$$\{U_1, U_2, \ldots, U_p\}$$

a cycle of Siegel discs. Then

$$\bigcup_{k=1}^{p} \partial(U_k) \subset S^+(f).$$

Proof. It suffices to consider an invariant Siegel disc U. By contradiction, suppose that there is a point $z_0 \in \partial(U)$ having a neighborhood V such that $V \cap S^+(f) = \emptyset$.

According to Theorem 7.4.2, f is univalent and bijective in U, and we can consider the inverse $h : U \to U$ of f. Obviously, we may extend each h^n to an analytic function on V. By Lemma 8.1.6, (h^n) is normal in V and, by Lemma 8.1.7, any limit function of a convergent subsequence of (h^n) is a constant. However, this is impossible since, by Theorem 7.4.2, there exists a subsequence converging to the identity on $U \cap V$. □

There are functions with a periodic cycle of Baker domains that do not contain any singular value (see, for example, [38]).

We leave the proof of the next theorem as Exercise 8.

Theorem 8.2.4. *If z_0 is a Cremer point of a transcendental entire function f, then $z_0 \in S^+(f)$.*

8.3 Special classes of entire functions

This section is purely informative, we only intend to point out some important properties of functions belonging to the widely studied classes $\mathcal{B}$ and $\mathcal{S}$, defined below. Among other results, we will see that, in some sense, properties of functions belonging to $\mathcal{S}$ resemble those of rational functions.

A transcendental entire function with the property that the set of singular values is bounded is called a function of *bounded singular type*. We denote by $\mathcal{B}$ the class of all those functions. Another widely studied class of functions is the class $\mathcal{S}$ (in honor of Speiser) formed by all entire functions that have only finitely many singular values. A function belonging to this class is called a function of *finite singular type*. The class $\mathcal{S}$ plays an important role in the value distribution theory and was investigated systematically by Nevanlinna, Teichmüller, and others, and it was introduced in the iteration theory in [37].

The next example shows that the classes $\mathcal{S}$ and $\mathcal{B}$ are different.

Example 8.3.1. The function $f(z) = \sin z/z$ has infinitely many singular values in the interval $[-1,1]$. The critical points of f are the roots of the equation $z\cos z - \sin z = 0$ or, equivalently, $z = \tan z$. If we denote by z_k the real roots, for $k \in \mathbb{Z}$, then $\sin^2 z_k = z_k^2/(1+z_k^2)$. Therefore, the real critical values of f are $f(z_k) = \pm 1/\sqrt{1+z_k^2}$ which belong to $[-1,1]$. As $z_k \to \infty$, it follows that $w_k = f(z_k) \to 0$, where each w_k is a critical value of f. Thus 0 is a singular value of f which is the limit of a sequence of singular values. The proof that $f \in \mathcal{B}$ is Exercise 10.

The next theorem is known as the second fundamental theorem for the class $\mathcal{S}$. For a proof, the reader may consult [63].

Theorem 8.3.2. *If $f \in \mathcal{S}$, then f has no wandering domains.*

Theorem 8.3.3. *Let $f \in \mathcal{S}$ be a transcendental entire function such that the orbit of every singular point is divergent or preperiodic. If $\mathbb{F}(f)$ has no superattracting cycles, then $\mathbb{J}(f) = \mathbb{C}$.*

Proof. As $\mathbb{F}(f)$ has no wandering domains, it suffices to prove that there are no periodic Fatou components. Let us see that $\mathbb{F}(f)$ has no attractive components. If U is an attractive fixed component and $z^* \in U$ is a fixed point of f that is not superattracting, then, by Theorem 8.2.1, there is a singular value z_0 of f in U and $f^n(z_0) \to z^*$. By hypothesis, the orbit of z_0 must be preperiodic and, in consequence, $O_f(z_0) = \{a, b, c, \ldots, z^*, z^*, \ldots\}$. This contradicts Corollary 5.1.3, which asserts that the orbit of z_0 should be infinite. Having in mind the remark after Theorem 5.3.6, by a similar argument, we may prove that $\mathbb{F}(f)$ has no parabolic fixed components. Finally, by Theorem 8.2.3, there are no Siegel discs either. □

Theorem 8.3.4. *Let $f \in \mathcal{B}$. For every $z \in \mathbb{F}(f)$, $(f^n(z))$ is not divergent.*

In particular, the latter theorem tells us that $\mathbb{F}(f)$ has no Baker domains for all $f \in \mathcal{B}$.

8.4 Permutable functions

In 1918, Fatou proved the following result (see Exercise 3):

"If f and g are permutable rational functions with degree at least two, then $\mathbb{F}(f) = \mathbb{F}(g)$."

In view of this result, he asked the following question: If f and g are permutable transcendental entire functions, does it follows that $\mathbb{F}(f) = \mathbb{F}(g)$? The question is still open. However, the answer is affirmative for certain classes of functions [58]. A result that is more related to singular values is the next one, due to Poon and Yang [47].

Theorem 8.4.1. *Let f and g be two permutable transcendental entire functions. If both $\operatorname{sing}(f^{-1})$ and $\operatorname{sing}(g^{-1})$ are isolated in the finite complex plane, then $\mathbb{F}(f) = \mathbb{F}(g)$.*

Proof. Obviously, it suffices to show that $\mathbb{F}(f) \subset \mathbb{F}(g)$. First, let us see that we only need to prove that $g(\mathbb{F}(f)) \subset \mathbb{F}(f)$. Indeed, if $g(\mathbb{F}(f)) \subset \mathbb{F}(f)$, then $g(U) \subset \mathbb{F}(f)$ for every component U of $\mathbb{F}(f)$ and, therefore, $g^n(U) \subset \mathbb{F}(f)$. Thus (g^n) omits at least two values in U and, in consequence, (g^n) is normal in U, which implies that $U \subset \mathbb{F}(g)$. Therefore, by contradiction, suppose there exist $z_0 \in \mathbb{J}(f)$ and $w_0 \in \mathbb{F}(f)$ such that $z_0 = g(w_0)$. We distinguish two cases:

(a) $z_0 \notin \operatorname{sing}(g^{-1})$. Then there is a disc $\mathbb{D}_r(z_0)$ with the property that an analytic branch g^{-1} of g is defined on $\mathbb{D}_r(z_0)$ with $g^{-1}(z_0) = w_0$. According to the first fundamental theorem, there exists a sequence of repelling periodic points (a_n) for f in $\mathbb{D}_r(z_0)$ such that $a_n \to z_0$ (denote the period of a_n by p_n). For each n, there is a disc $\mathbb{D}_{r_n}(a_n) \subset \mathbb{D}_r(z_0)$. As $f^{p_n}(a_n) = a_n$, by continuity of f^{p_n}, there exists a disc $\mathbb{D}_{s_n}(a_n)$ such that $f^{p_n}(\mathbb{D}_{s_n}(a_n)) \subset \mathbb{D}_{r_n}(a_n)$. Then, for every $z \in \mathbb{D}_{s_n}(a_n)$,

$$g^{-1} \circ f^{p_n}(z) = f^{p_n} \circ g^{-1}(z). \tag{8.6}$$

Indeed, $g(f^{p_n} \circ g^{-1}(z)) = f^{p_n}(z)$ since f and g are permutable.

Taking the derivative on both sides of (8.6), we obtain

$$(g^{-1})'(a_n)(f^{p_n})'(a_n) = (f^{p_n})'(g^{-1}(a_n))(g^{-1})'(a_n).$$

This and (8.6) imply that $g^{-1}(a_n)$ is a repelling periodic point for f. On the other hand, $g^{-1}(a_n) \to g^{-1}(z_0) = w_0$. Since $\mathbb{J}(f)$ is closed, we conclude that $w_0 \in \mathbb{J}(f)$, and this is a contradiction.

(b) $z_0 \in \operatorname{sing}(g^{-1})$. By hypothesis, there is a disc $\mathbb{D}_r(z_0)$ such that z_0 is the only singular value of g in $\mathbb{D}_r(z_0)$. Since $z_0 = g(w_0)$ and $w_0 \in \mathbb{F}(f)$, we can choose a disc $U_0 \subset \mathbb{F}(f)$ such that $w_0 \in U_0$ and $g(U_0) \subset \mathbb{D}_r(z_0)$. Now we show that z_0 is the only point in $g(U_0)$ that belongs to $\mathbb{J}(f)$. Indeed, suppose that $z_1 \in (g(U_0) \setminus \{z_0\}) \cap \mathbb{J}(f)$. Then $z_1 = g(w_1)$, where $w_1 \in U_0 \subset \mathbb{F}(f)$. Notice that $z_1 \in \mathbb{D}_r(z_0)$ and, therefore, z_1 is not a singular value of g. Then the pair (w_1, z_1) satisfies the conditions of case (a), which is impossible. Hence $g(U_0)$ is a neighborhood of z_0 that has only the point z_0 in common with $\mathbb{J}(f)$. This is a contradiction since the Julia set is perfect. □

8.5 Completely invariant components

Recall that a Fatou component U is called completely invariant if $f(U) \subset U$ and $f^{-1}(U) \subset U$.

Theorem 8.5.1. *If f is a transcendental entire function and U is a completely invariant Fatou component, then*
(i) *U is unbounded.*
(ii) *Every Fatou component is simply connected.*
(iii) $\mathbb{J}(f) = \partial(U)$.

Proof. (i) By the Picard theorem, the equation $f(z) = a$ has infinitely many solutions for all $a \in U$, except for at most a point. As U is completely invariant, all these solutions belong to U. Then take some $a \in U$ such that the set $A = \{z \in U : f(z) = a\}$ is infinite. If U is bounded, then A is relatively compact and, therefore, there is a sequence (z_n) in A which is convergent to z_0. By continuity, $f(z_0) = a$. Thus the identity principle implies that $f(z) \equiv a$.

(ii) As U is unbounded, it follows from Corollary 7.2.5 that every Fatou component is simply connected.

(iii) Suppose that there is a $z_0 \in \mathbb{J}(f)$ such that $z_0 \notin \partial(U)$. Then there exists a disc $\mathbb{D}(z_0)$ such that either $\mathbb{D}(z_0) \subset U$ or $\mathbb{D}(z_0) \subset \mathbb{C} \setminus U$. In the former case, we have $\mathbb{D}(z_0) \subset \mathbb{F}(f)$ and, in the latter, $f^n(\mathbb{D}(z_0)) \cap U = \emptyset$ for all n since U is completely invariant. This shows that (f^n) is normal in $\mathbb{D}(z_0)$, which is a contradiction. □

In 1970, Baker proved that a transcendental entire function has at most one completely invariant Fatou component [7]. The following open conjecture is due to Baker [8].

Conjecture 8.5.2. *Let f be a transcendental entire function. If $\mathbb{F}(f)$ has a completely invariant component U, then $\mathbb{F}(f) = U$.*

The conjecture is true for some classes of functions. We finish the section with a result of this type (see [47]).

Theorem 8.5.3. *Let f a transcendental entire function in the class $\mathcal{B}$ such that $\mathbb{F}(f)$ has a completely invariant component U. If $\mathbb{J}(f)$ does not contain any limit point of $\bigcup_{n\geq 1} f^n(\mathrm{sing}(f^{-1}))$, then $\mathbb{F}(f) = U$.*

8.6 Connectivity of Julia sets

In the framework of polynomial functions, the following theorem about the connectivity of the Julia set is well known (see, for example, [13, Section 9.5]). We will denote by $\mathbb{F}_\infty(P)$ the unbounded Fatou component of $\mathbb{F}(P)$ whenever P is a polynomial with degree $d \geq 2$.

Theorem 8.6.1. *If P is a polynomial with degree $d \geq 2$, then the following statements are equivalent:*
(i) *$\mathbb{F}_\infty(P)$ is simply connected.*
(ii) *$\mathbb{J}(P)$ is connected.*
(iii) *There are no finite critical points of P in $\mathbb{F}_\infty(P)$.*

Nevertheless, the latter theorem does not hold for transcendental entire functions.

Example 8.6.2. Consider the function $E_\lambda(z) = \lambda e^z$ for $0 < \lambda < 1/e$. In the next chapter, we will see that E_λ has an attracting fixed point q_λ and its basin contains the origin. Then there are no finite singular values with a divergent orbit. However, we will see in Chapter 9 that $\mathbb{J}(E_\lambda)$ is disconnected.

We need the following standard results on simply connected domains [67].

Proposition 8.6.3. (1) *Let $K \subset \hat{\mathbb{C}}$ be compact. Then K is disconnected if and only if there exists a Jordan curve γ which separates K.*
(2) *If U is a domain, the following statements are equivalent:*
 (i) *U is simply connected.*
 (ii) *$\hat{\mathbb{C}} \setminus U$ is connected.*
 (iii) *$\partial(U)$ is connected.*

With this proposition in hand, we may prove the next result.

Proposition 8.6.4. *Let $K \subset \hat{\mathbb{C}}$ compact. Then K is connected if and only each component of the complement K^c is simply connected.*

Proof. ($\Longrightarrow$) If K is connected, consider a component D of K^c and denote by V the complement of D. It is not difficult to see that V is connected and, therefore, D is simply connected.

($\Longleftarrow$) By contradiction, suppose that K is disconnected. By Proposition 8.6.3, there is a Jordan curve γ which separates K. The curve γ must lie in some component D of K^C and then D cannot be simply connected. □

Corollary 8.6.5. *Let f be a transcendental entire function. Then $\mathbb{J}(f) \cup \{\infty\}$ is connected if and only if each component of $\mathbb{F}(f)$ is simply connected.*

Functions in the class $\mathcal{B}$ have the property that every Fatou component is simply connected. So, we have proved the next corollary.

Corollary 8.6.6. *If f is a transcendental entire function belonging to the class $\mathcal{B}$, then $\mathbb{J}(f) \cup \{\infty\}$ is connected.*

Finally, we establish a sufficient condition for $\mathbb{J}(f)$ to be connected.

Corollary 8.6.7. *Let f be a transcendental entire function. If every Fatou component is simply connected and bounded, then $\mathbb{J}(f)$ is connected.*

8.7 Exercises

1. Consider the entire function defined by $F(z) = \int_0^z \frac{\sin w}{w} dw$. Prove that $z_0 = \int_0^\infty \frac{\sin r}{r} dr$ is an asymptotic value of F.
2. If $f(z) = z + e^z + \log c$, where c is a positive real number, show that the singular values of f are the points $z_n = -1 + \log c + (2n+1)\pi i$ for every $n \in \mathbb{Z}$.
3. Let a be a complex number such that $|a| < 1$. Show that the functions $f(z) = az$ and $g(z) = z/a$ are permutable, but have different Julia sets.
4. Prove that the functions $\lambda \sin z$ and $\lambda \cos z$ have only two critical values and no (finite) asymptotic values.
5. Let $f(z) = e^{-e^z}$. Show that f has only two asymptotic values, namely 0 and 1.
6. Let λ be a positive real number sufficiently small and let $n \geq 2$ be a positive integer. If $f(z) = \lambda e^{z^n}$, prove the following statements:
   (i) The function f has only two singular values, namely λ and 0.
   (ii) The function f has a unique attracting fixed point z^*.
   (iii) $\mathbb{F}(f) = \Omega(f, z^*)$.
7. Let $f(z) = \lambda \cos^2 z$ with $\lambda \in (0, 1)$. Show that:
   (i) The function f has only one real fixed point x_0 (attracting).
   (ii) $S^+(f) = O_f^+(0) \cup \{x_0\}$.
   (iii) $S^+(f) \subset \Omega_0(f, x_0)$.
   (iv) The set $\Omega_0(f, x_0)$ is the unique periodic component of $\mathbb{F}(f)$.
8. Let $f \in \mathcal{S}$ be with an attracting fixed point z^* such that $S^+(f) \subset \Omega_0(f, z^*)$. Show that $\mathbb{F}(f) = \Omega(f, z^*)$.
9. Prove that every Cremer point belongs to $S^+(f)$.
10. Consider the function $f(z) = \sin z/z$.

(i) Show that the origin is the unique asymptotic value of f.
(ii) Prove that f belongs to the class $\mathcal{B}$.

11. Suppose f and g are permutable entire functions. If U is a bounded periodic component of $\mathbb{F}(f)$, then $U \subset \mathbb{F}(g)$.
12. Describe the dynamics of $f(z) = z^2 - z^3$.
13. Under the conditions of Theorem 8.1.3(2b), prove that $f \circ \gamma_j(t) \to c$, with c being a certain solution of the equation $f^{n-j-1}(z) = a$.

Bibliography remarks

Section 8.2 is based on [61, Section 8] and, for the Mittag-Leffler star, we followed [60, Section 8.5].

In Section 8.4, we follow [47, Section 7.2].

9 The exponential family

The dynamical properties of the *exponential family* $E_\lambda(z) = \lambda e^z$ ($\lambda \in \mathbb{C}$) are well known, especially in the case the parameter λ is real [31]. In 1981, Misiurewicz proved that the Julia set of $E(z) = e^z$ is $\mathbb{C}$, answering a sixty-year-old question of Fatou. The main purpose of this chapter is to describe the Julia set of E_λ. According to Corollary 7.2.6, if f is a transcendental entire function, the Julia set of f is not totally disconnected. However, it is known that there are meromorphic functions such that $\mathbb{J}(f)$ is a Cantor set (see, for instance, [30]). We will show that the Julia set of $E_\lambda(z) = \lambda e^z$, for every $\lambda \in (0, 1/e)$, contains an interesting and crazy set, namely, a Cantor bouquet. Roughly speaking, a Cantor bouquet is an uncountable collection of disjoint continuous curves tending to ∞ in a certain direction on the plane, each of which has a distinguished endpoint. We will make this notion precise later.

This chapter has been elaborated following mainly [63] and [32], and with plenty of personal work.

9.1 Periodic points

We start studying the existence of real fixed points of E_λ for $\lambda \in \mathbb{R}$.

Proposition 9.1.1. *If $\lambda \in \mathbb{R}$, then the exponential family has the following real fixed points according to the value of λ:*

(a) *If $\lambda = 1/e$, $x^* = 1$ is a neutral fixed point and it is the unique real fixed point.*

(b) *If $\lambda > 1/e$, E_λ has no real fixed points.*

(c) *If $0 < \lambda < 1/e$, E_λ has two fixed points, namely $x^* \in (0, 1)$ (attracting) and $x^{**} > 1$ (repelling).*

(d) *If $\lambda < 0$, E_λ has a unique real fixed point x^* such that*
   - *x^* is attracting if $-e < \lambda < 0$.*
   - *x^* is repelling if $\lambda < -e$.*
   - *x^* is neutral if $\lambda = -e$.*

Proof. (a) $\lambda = 1/e$. Obviously, $x^* = 1$ is a real fixed point. As $E_\lambda'(x^*) = \lambda e^{x^*} = x^* = 1$, this fixed point is neutral. Also $y = e^x$ and $y = ex$ are tangent at $x = 1$, that is why there are no other real fixed points.

(b) $\lambda > 1/e$. The equation $f(x) = \lambda e^x - x = 0$ has no real solutions. Since $\lambda e^x - x > 0$ for $x < 0$, it is obvious that there are no negative solutions. Thus we study the sign of the derivative for $x > 0$:

$$f'(x) = \lambda\left(e^x - \frac{1}{\lambda}\right)\begin{cases} < 0 & \text{if } 0 \le x \le \log(1/\lambda), \\ > 0 & \text{if } \log(1/\lambda) \le x. \end{cases}$$

https://doi.org/10.1515/9783111689685-009

We see that $x_0 = \log(1/\lambda)$ is an absolute minimum of f in $[0, +\infty)$, thus $f(x) \geq f(x_0) = 1 - \log(1/\lambda) > 0$, for all $x \geq 0$.

(c) $0 < \lambda < 1/e$. Again, it is obvious that there are no negative real fixed points and a simple application of Bolzano's theorem shows that there exist two positive fixed points, namely $x^* \in (0,1)$ (attracting) and $x^{**} > 1$ (repelling).

(d) $\lambda < 0$. In this case, $f'(x) = \lambda e^x - 1 < 0$ for all $x \in \mathbb{R}$. Then f is strictly decreasing in $\mathbb{R}$. In view of the limits at the infinity,

$$\lim_{x\to-\infty} f(x) = +\infty \quad \text{and} \quad \lim_{x\to+\infty} f(x) = -\infty,$$

we deduce that there exists a unique fixed point x^* for E_λ. The following possibilities may occur:

- $-e < \lambda < 0$. In this case, $f(0) = \lambda < 0$ and $f(-1) = \lambda/e + 1 > 0$; thus $x^* \in (-1, 0)$ and it is attracting.
- $\lambda < -e$. Now $f(-1) = \lambda/e + 1 < 0$, but $\lim_{x\to+\infty} f(x) = +\infty$. Thus $x^* < -1$ and it is repelling.
- $\lambda = -e$. In this case, $x^* = -1$ is the unique fixed point of E_λ and it is neutral. □

As we will see in the final section, there are values of λ for which $\mathbb{J}(E_\lambda) = \mathbb{C}$. In view of the above proposition, the family $E_\lambda = \lambda e^z$ for $\lambda \in (0, 1/e)$ is the most interesting case and has been widely studied. We denote by q_λ and p_λ the attracting and repelling points, respectively, of E_λ for $\lambda \in (0, 1/e)$.

Lemma 9.1.2. *For each $\lambda \in (0, 1/e)$, the following statements hold:*
(a) $q_\lambda < \log(1/\lambda) < p_\lambda$.
(b) *If ℓ_λ is a real number in the interval $(\log(1/\lambda), p_\lambda)$, then $E_\lambda(\ell_\lambda) < \ell_\lambda$.*

Proof. Since q_λ and p_λ are fixed points of E_λ, it follows that

$$q_\lambda = E_\lambda(q_\lambda) < 1 = E_\lambda(\log(1/\lambda)) < E_\lambda(p_\lambda) = p_\lambda.$$

But the above relation is equivalent to

$$\lambda e^{q_\lambda} < \lambda e^{\log(1/\lambda)} < \lambda e^{p_\lambda},$$

which yields (a).

To prove (b), consider the function $g(x) = e^x/x$ for $x \geq 1$. Notice that $g'(x) > 0$ for $x > 1$, then g is strictly increasing in $[1, +\infty)$. As $g(1) = e$, $g(p_\lambda) = 1/\lambda$, and $1 < \log(1/\lambda) < p_\lambda$, choosing ℓ_λ in the interval $(\log(1/\lambda), p_\lambda)$, we have $g(\ell_\lambda) < g(p_\lambda) = 1/\lambda$. Therefore, with this choice of ℓ_λ, the proof concludes. □

The next proposition lists some results that we will use throughout the chapter.

Proposition 9.1.3. *Let $\lambda \in (0, 1/e)$ and let ℓ_λ be a real number in the interval $(\log(1/\lambda), p_\lambda)$. The following statements hold:*

(a) *The function E_λ maps the half-plane $\mathbb{R}e\, z < \ell_\lambda$ to itself.*
(b) *The half-plane $\mathbb{R}e\, z < p_\lambda$ is contained in the attractive basin of the fixed point q_λ.*
(c) *There exists $\mu > 1$ such that $|E'_\lambda(z)| \geq \mu$ for all z in the half-plane $\mathbb{R}e\, z \geq \ell_\lambda$.*
(d) *$|E'_\lambda(z)| > 1$ in the half-plane $\mathbb{R}e(z) \geq \log(1/\lambda)$.*
(e) *Let $v = \log(1/\lambda) > 1$ and denote by H the half-plane $\mathbb{R}e(z) < v$. Then H is contained in the immediate basin of attraction of q_λ.*
(f) *The horizontal line $\mathbb{I}m(z) = (2n+1)\pi$ is contained in*

$$E_\lambda^{-1}(\Omega_0(E_\lambda, q_\lambda)) \cap \Omega_0(E_\lambda, q_\lambda),$$

for every $n \in \mathbb{Z}$.

Proof. (a) For a z satisfying $\mathbb{R}e\, z < \ell_\lambda$, by Lemma 9.1.2(b), we have

$$\mathbb{R}e(E_\lambda(z)) = \lambda e^{\mathbb{R}e\, z} \cos(\mathbb{I}m\, z) < \lambda e^{\ell_\lambda} = E_\lambda(\ell_\lambda) < \ell_\lambda.$$

(b) Given a z such that $\mathbb{R}e(z) < p_\lambda$, we choose $\ell_\lambda \in (\log(1/\lambda), p_\lambda) \cap (\mathbb{R}e(z), p_\lambda)$. Proposition 4.2.3 applied to the half-plane $\mathbb{R}e(\zeta) < \ell_\lambda$ tells us that the orbits starting in the half-plane $\mathbb{R}e(\zeta) < \ell_\lambda$ are convergent to q_λ.

(c) If $\mathbb{R}e\, z \geq \ell_\lambda$, we have

$$|E'_\lambda(z)| = \lambda e^{\mathbb{R}e\, z} \geq \lambda e^{\ell_\lambda} = \mu > \lambda e^{\log(1/\lambda)} = 1.$$

(d) The argument is similar to that used in the proof of (b).

(e) Note that E_λ maps H onto $\mathbb{D}^*$ and then $E_\lambda(H) \subset H$. According to Proposition 4.2.3, H is contained in the immediate basin of attraction of q_λ.

(f) If $z = x + (2n+1)\pi i$, then $E_\lambda(z) = -\lambda e^x \in H \subset \Omega_0(E_\lambda, q_\lambda)$. On the other hand, as the attractive basin is completely invariant, it follows that $z \in \Omega(E_\lambda, q_\lambda)$. So, as the set $H \cup \{z : \mathbb{I}m(z) = (2n+1)\pi\}$ is connected, we conclude that $\{z : \mathbb{I}m(z) = (2n+1)\pi\}$ is contained in $\Omega_0(E_\lambda, q_\lambda)$. □

Corollary 9.1.4. *If $\lambda \in (0, 1/e)$, then the Julia set of E_λ is contained in the half-plane $\mathbb{R}e\, z \geq p_\lambda$.*

Corollary 9.1.5. *If $\lambda \in (0, 1/e)$, then E_λ has no neutral fixed point and q_λ is the only attracting fixed point.*

Proof. If z_0 is a point such that $|(E_\lambda)'(z_0)| \leq 1$, then $\mathbb{R}e(z_0) \leq \log(1/\lambda)$. So, z_0 belongs to the half-plane $\mathbb{R}e(z) < p_\lambda$ and Proposition 9.1.3 tells us that z_0 is in the attractive basin of q_λ. □

Corollary 9.1.6. *If $\lambda \in (0, 1/e)$ and $p > 1$, then every p-periodic point of E_λ is repelling.*

Proof. Let z_0 be a periodic point of E_λ with period p and $z_0, z_1, \ldots, z_{p-1}$ the correspondent p-cycle. Each z_j has a periodic orbit and, in consequence, it is not convergent. Thus none of the z_j's belong to the half-plane $\mathbb{R}e(\zeta) < p_\lambda$. Then $\mathbb{R}e(z_j) \geq p_\lambda > \log(1/\lambda)$ and, according Proposition 9.1.3(c), we can conclude that $|E_\lambda'(z_j)| > 1$ for $j = 0, 1, \ldots, p-1$. Finally, by the chain rule,

$$(E_\lambda^p)'(z_0) = \prod_{j=0}^{p-1} E_\lambda'(z_j),$$

which yields $|(E_\lambda^p)'(z_0)| > 1$. □

The next proposition is due to Devaney and Tangerman [32] (Exercise 1).

Proposition 9.1.7. *The attractive basin of q_λ is dense in $\mathbb{C}$.*

We finish this section proving that the immediate basin of attraction of q_λ coincides with the Fatou set of E_λ.

Theorem 9.1.8. *If $\lambda \in (0, 1/e)$, then $\Omega_0(E_\lambda, q_\lambda) = \mathbb{F}(E_\lambda)$.*

Proof. As the origin is the only singular value of E_λ and the orbit of this point is convergent to q_λ, combining Theorems 8.2.1, 8.2.2, 8.2.3, and 8.3.2, we deduce that $\Omega_0(E_\lambda, q_\lambda)$ is the unique periodic component of $\mathbb{F}(E_\lambda)$ and, in consequence, $\mathbb{F}(E_\lambda) = \Omega(E_\lambda, q_\lambda)$. On the other hand, notice that $H = E_\lambda^{-1}(\mathbb{D})$ and, therefore, we have $H \subset E_\lambda^{-1}(\Omega_0(E_\lambda, q_\lambda))$. So, if we show that $E_\lambda^{-1}(\Omega_0(E_\lambda, q_\lambda))$ is connected, we conclude that $E_\lambda^{-1}(\Omega_0(E_\lambda, q_\lambda)) \subset \Omega_0(E_\lambda, q_\lambda)$. For this, in view of Proposition 9.1.3(f), it suffices to prove that we can connect (with a curve in $E_\lambda^{-1}(\Omega_0(E_\lambda, q_\lambda)))$ two points $z_1, z_2 \in E_\lambda^{-1}(\Omega_0(E_\lambda, q_\lambda))$ such that $z_j \notin H$ $(j = 1, 2)$ and $(2n-1)\pi < \mathbb{I}m(z_j) < (2n+1)\pi$ for $j = 1, 2$ and some $n \in \mathbb{Z}$. Put $w_j = E_\lambda(z_j)$ and note that $w_j \in \Omega_0(E_\lambda, q_\lambda) \setminus \mathbb{D}$. It is easy to show that w_1 and w_2 may be connected with a curve γ in $\Omega_0(E_\lambda, q_\lambda)$ which does not intersect $\mathbb{R}^- = \{z = x + 0i : x \leq 0\}$. Then we can consider the inverse L_λ of E_λ defined by

$$L_\lambda(w) = \log\left(\frac{|w|}{\lambda}\right) + \arg(w)i \quad \text{for } w \in \mathbb{C} \setminus \mathbb{R}^-,$$

where $\arg(w) \in ((2n-1)\pi, (2n+1)\pi)$. Therefore, $L_\lambda \circ \gamma$ is a curve in $E_\lambda^{-1}(\Omega_0(E_\lambda, q_\lambda))$ connecting the points z_1 and z_2. Summing up, we have proved that $\Omega_0(E_\lambda, q_\lambda)$ is a completely invariant component of $\mathbb{F}(E_\lambda)$ and Theorem 8.5.3 tells us that it is the only Fatou component of E_λ. □

9.2 Symbolic dynamics

In the next sections, we will use symbolic dynamics techniques to describe the Julia set of E_λ. Symbolic dynamics is a tool to analyze general dynamical systems by discretizing space. Imagine a point following some trajectory in a space. We consider a partition of the space into finitely many pieces, each labeled with a different symbol. Writing down the sequence of symbols corresponding to the successive partition elements visited by the point in its orbit, we obtain a symbolic trajectory. We may ask: Can we learn anything about the dynamics of the system by scrutinizing its symbolic trajectory?

Hadamard is generally credited with the first successful use of symbolic dynamics in his analysis of geodesic flows on surfaces of negative curvature in 1898. Forty years later, the subject received its first systematic study in the fundamental paper of M. Morse and G. Hedlung [64]. This study was motivated both by the intrinsic mathematical interest in symbolic dynamics and the need to better understand it in order to apply symbolic techniques to continuous systems.

Given a positive integer $N \geq 2$, we denote by Σ_N the set of infinite sequences $\overline{s} = (s_0, s_1, s_2, \dots)$ defined by

$$\Sigma_N = \{\overline{s} = (s_0, s_1, s_2, \dots) : -N \leq s_i \leq N, s_i \in \mathbb{Z}\}.$$

The set Σ_N is endowed with the metric

$$d(\overline{s}, \overline{s}^*) = \sum_{i=0}^{\infty} \frac{|s_i - s_i^*|}{N^i}.$$

The next lemma allows us to prove easily that, for each $\overline{s} \in \Sigma_N$, the sets

$$V(\overline{s}, k) = \{\overline{t} = (t_i) \in \Sigma_N : t_i = s_i \; i = 0, 1, \dots, k\},$$

for $k = 0, 1, 2, \dots$, form a neighborhood basis of $\overline{s}$ for the metric topology and, in consequence, this topology coincides with the product topology for Σ_N.

Lemma 9.2.1. (i) *For all $\overline{s}, \overline{t} \in \Sigma_N$, if $s_i = t_i$ for $i \leq k$, then $d(\overline{s}, \overline{t}) \leq 2/N^{k-1}(N-1)$.*
(ii) *If $d(\overline{s}, \overline{t}) < 1/N^k$, then $s_i = t_i$ for $i \leq k$.*

Proof. (i) Let $\overline{s}, \overline{t} \in \Sigma_N$ be such that $s_i = t_i$ for $i \leq k$. Then we have

$$d(\overline{s}, \overline{t}) = \sum_{i \geq k+1} \frac{|s_i - t_i|}{N^i} \leq 2N \sum_{i \geq k+1} \frac{1}{N^i} = \frac{2}{N^{k-1}(N-1)}.$$

(ii) If $s_i \neq t_i$ for some $i \leq k$, then $d(\overline{s}, \overline{t}) \geq |s_i - t_i|/N^i > 1/N^i \geq 1/N^k$. □

Recall that a topological space X is perfect if every point is a limit point or, equivalently, it does not contain any isolated point. We say that X is totally disconnected if every connected subset of X is a singleton.

Definition 9.2.2. Let X be a metric space. We say that X is a Cantor set if it is compact, perfect, and totally disconnected.

The most famous example is the Cantor middle-third set.

Proposition 9.2.3. *The set Σ_N is a Cantor set.*

Proof. (1) The set Σ_N is perfect. Given $\overline{s}$ and $\varepsilon > 0$, choose $k \in \mathbb{N}$ such that $2/N^{k-1}(N-1) < \varepsilon$. By Lemma 9.2.1, if $\overline{t} \in \Sigma_N$ is such that $t_i = s_i$ for $i \leq k$, then

$$d(\overline{s},\overline{t}) \le \frac{2}{N^{k-1}(N-1)} < \varepsilon,$$

which implies that $B(\overline{s},\varepsilon)$ contains infinitely many elements.

(2) The set Σ_N is totally disconnected. Otherwise, suppose that $X \subset \Sigma_N$ is connected and $\overline{s}, \overline{s}^* \in \Sigma_N$ with $\overline{s} \neq \overline{s}^*$. Then there is an $i \in \mathbb{N}$ such that $s_i \neq s_i^*$. We consider the sets

$$X_1 = \{\overline{t} \in \Sigma_N : t_i = s_i\} \quad \text{and} \quad X_2 = \{\overline{t} \in \Sigma_N : t_i \neq s_i\}.$$

Notice that $(X_1 \cap X) \cup (X_2 \cap X) = X$, $X_1 \cap X_2 = \emptyset$, $\overline{s} \in X_1$, and $\overline{s}^* \in X_2$. Therefore, if we prove that X_1 and X_2 are open, we will reach a contradiction. As $X_2 = \bigcup_{s \neq s_i} \{\overline{t} \in \Sigma_N : t_i = s\}$, it suffices to show that X_1 is open. For it, given $\overline{t} \in X_1$, choose $\varepsilon > 0$ such that $\varepsilon < 1/N^i$. By Lemma 9.2.1, $\overline{r} \in B(\overline{t},\varepsilon)$ implies that $r_j = t_j$ for $j \le i$. In particular, we have $r_i = s_i$ and, therefore, $B(\overline{t},\varepsilon) \subset X_1$.

(3) According to Tychonoff's theorem, Σ_N is compact. □

9.3 The Julia set of E_λ for $0 < \lambda < 1/e$

Throughout this section, we suppose that $0 < \lambda < 1/e$. We have just seen that E_λ has two real fixed points, namely q_λ (attracting) and p_λ (repelling) such that $0 < q_\lambda < 1 < p_\lambda$.

For fixed $\lambda \in (0, 1/e)$ and $\ell_\lambda \in (\log(1/\lambda), p_\lambda)$, we will construct in the half-plane $\mathbb{R}e\, z \ge \ell_\lambda$ a collection of invariant Cantor sets for E_λ.

Given $N \in \mathbb{N}$, denote by $B(N)$ the rectangle formed by all $z = x + yi$ such that

$$x \in [\ell_\lambda, r_\lambda] \quad \text{and} \quad y \in [-(2N+1)\pi, (2N+1)\pi],$$

where r_λ is such that $\lambda e^{r_\lambda} > r_\lambda + (2N+1)\pi$ (see Figure 9.1).

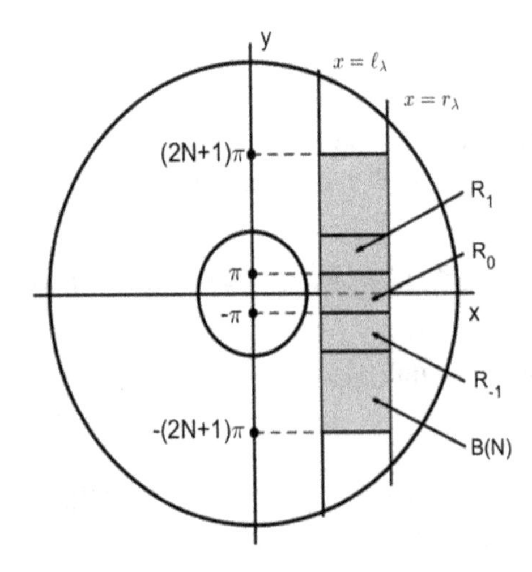

Figure 9.1: $B(N)$ is the shaded rectangle.

The function E_λ maps the right vertical boundary of $B(N)$ to the circle of radius λe^{r_λ} with center the origin, while the left one is mapped to the circle of radius λe^{ℓ_λ}. Consider the ring $A = \{z : \lambda e^{\ell_\lambda} \le |z| \le \lambda e^{r_\lambda}\}$. We will prove the following properties:

(1) One has $B(N) \subset \mathring{A}$. If $z = x + iy \in B(N)$, we get

$$\ell_\lambda \le x \le r_\lambda, \quad -(2N+1)\pi \le y \le (2N+1)\pi.$$

Therefore

$$|z| \le |x| + |y| \le r_\lambda + (2N+1)\pi < \lambda e^{r_\lambda}.$$

On the other hand, if $z \in B(N)$, obviously $|z| \ge \ell_\lambda > \lambda e^{\ell_\lambda}$.

(2) The function E_λ maps $B(N)$ onto A. If $z \in B(N)$, then $E_\lambda(z) = \lambda e^x e^{yi}$ with x and y as in (1). Thus

$$|E_\lambda(z)| = \lambda e^x \le \lambda e^{r_\lambda} \quad \text{and} \quad |E_\lambda(z)| = \lambda e^x \ge \lambda e^{\ell_\lambda}.$$

Now, for each $i = -N, \dots, N$, let $R_i \subset B(N)$ be the subrectangle

$$\ell_\lambda \le \mathbb{R}\mathrm{e}\, z \le r_\lambda, \quad (2i-1)\pi \le \mathbb{I}\mathrm{m}\, z \le (2i+1)\pi.$$

By the same argument as that used in (2), we can prove that E_λ maps R_i to the ring A. For the surjectivity, take $w = r\lambda e^{i\theta} \in A$, with $r \in (e^{\ell_\lambda}, e^{r_\lambda})$ and $\theta \in [(2i-1)\pi, (2i+1)\pi]$. Now put $x = \log r$, $z = x + \theta i$, and note that $z \in R_i$ and $E_\lambda(z) = \lambda e^x e^{\theta i} = \lambda r e^{\theta i} = w$. From the equality $E_\lambda(R_i) = A$, it follows that $R_j \subset E_\lambda(R_i)$ for any i, j.

By Proposition 9.1.3(c), there exists a $\mu > 1$ such that

$$|E'_\lambda(z)| \ge \mu > 1 \quad \text{for all } \mathbb{R}\mathrm{e}(z) \ge \ell_\lambda. \tag{9.1}$$

Then the above property holds in each R_i.

Now define Λ_N by

$$\Lambda_N = \{z \in B(N) : E^j_\lambda(z) \in B(N)\ (\forall j)\}. \tag{9.2}$$

For each $k = 0, \pm 1, \pm 2, \dots, \pm N$, put

$$S(k) = \{z : (2\,k-1)\pi \le \mathbb{I}\mathrm{m}(z) \le (2k+1)\pi\}.$$

We denote by L_k the branch of the inverse of E_λ defined on the complex plane without the semiaxis $\{x + 0i : x \le 0\}$ and with values in $S(k)$. Then, if $z \in \mathbb{C} \setminus \mathbb{R}^-$, $L_k(z) = \log(|z|/\lambda) + \arg(z)i$, with $\arg(z) \in [(2k-1)\pi, (2k+1)\pi]$. For every $k = -N, \dots, N$, we have

$$E_\lambda \circ L_k(z) = z \quad \text{for all } z \in \mathbb{C} \setminus \{z = x + 0i : x \le 0\}. \tag{9.3}$$

Claim 1. $L_k(R_j) \subset \mathring{R}_k$ for all j, k.

As $R_j \subset A$, each $w \in R_j$ can be expressed in the form $w = r\lambda e^{\theta i}$, with $r \in (e^{\ell_\lambda}, e^{r_\lambda})$ and $\theta \in ((2k-1)\pi, (2k+1)\pi)$. Then $L_k(w) = \log r + i\theta$, which implies $L_k(w) \in \mathring{R}_k$ since r belongs to $(e^{\ell_\lambda}, e^{r_\lambda})$.

Given $z \in \Lambda_N$, note that $\mathbb{I}m(E_\lambda^n(z)) \neq (2j \pm 1)\pi$ for all nonnegative integers n and $j = -N, \ldots, N$. Indeed, if $E_\lambda^n(z) = x + (2j \pm 1)\pi i$, then $E_\lambda^{n+1}(z) = -\lambda e^x$, which is not in $B(N)$. Then each $E_\lambda^n(z)$ is contained in a concrete R_i^* (R_i^* is the subrectangle R_i without the horizontal boundaries). In consequence, each $z \in \Lambda_N$ determines a unique sequence $\overline{s} = (s_0, s_1, s_2, \ldots)$, with $s_i = -N, \ldots, N$ for all i, such that

$$E_\lambda^n(z) \in R_{s_n}^* \quad \text{for all } n = 0, 1, \ldots \tag{9.4}$$

The sequence $s_0, s_1, s_2, \ldots$ is called the *itinerary* of z.

Under the above conditions, we have the following equality:

$$Ls_n \circ E_\lambda(E_\lambda^n(z)) = E_\lambda^n(z). \tag{9.5}$$

To see this, put $w = E_\lambda^n(z)$. We know that $w \in R_{s_n}^*$, thus $w = x + yi$ with $x \in [\ell_\lambda, r_\lambda]$ and $y \in ((2s_n - 1)\pi, (2s_n + 1)\pi)$. Therefore, $L_{s_n}(E_\lambda(w)) = L_{s_n}(\lambda e^{x+iy}) = x + iy$.

Now, for each $\overline{s} \in \Sigma_N$, we prove that the set

$$\bigcap_{n \geq 0} L_{s_0} \circ \cdots \circ L_{s_{n-1}}(B(N))$$

contains only one point that we denote by $z_{\overline{s}}$. Observe that $(L_{s_0} \circ \cdots \circ L_{s_{n-1}}(B(N)))_n$ is a decreasing sequence of nonempty compact sets whose diameters d_n tend to 0. Indeed, using Claim 1 and induction, we can prove that the sequence is decreasing. Let us show that (d_n) converges to 0. For $z_2, z_1 \in B(N)$, applying (9.1), (9.3), and the chain rule, we obtain

$$\begin{aligned}
&|L_{s_0} \circ \cdots \circ L_{s_{n-1}}(z_2) - L_{s_0} \circ \cdots \circ L_{s_{n-1}}(z_1)| \\
&\quad = \left| \int_{z_1}^{z_2} (L_{s_0} \circ \cdots \circ L_{s_{n-1}})'(z)\, dz \right| \\
&\quad \leq |z_2 - z_1| \sup\{|(L_{s_0} \circ \cdots \circ L_{s_{n-1}})'(z)| : z \in B(N)\} \leq \frac{d(B(N))}{\mu^{n+1}}.
\end{aligned}$$

Then we may define the map

$$\phi : \overline{s} \in \Sigma_N \to \phi(\overline{s}) = z_{\overline{s}} \in \Lambda_N. \tag{9.6}$$

We recall that the shift map is defined from Σ_N into itself by

$$\sigma(s_0, s_1, s_2, \ldots) = (s_1, s_2, \ldots)$$

and it is continuous.

Theorem 9.3.1. *The map ϕ is a homeomorphism and the restriction of E_λ to Λ_N is topologically conjugate to σ through ϕ.*

Proof. 1. The map ϕ is surjective. For a fixed $z \in \Lambda_N$, let $\overline{s} = (s_n) \in \Sigma_N$ be the sequence given by (9.4). We will prove that $\phi(\overline{s}) = z$. Obviously, $L_{s_0} \circ E_\lambda(z) = z$. Reasoning by induction, suppose that

$$z = L_{s_0} \circ \cdots \circ L_{s_n}(E_\lambda^{n+1}(z)).$$

Using the induction hypothesis and (9.5), we obtain

$$\begin{aligned} &L_{s_0} \circ L_{s_1} \circ \cdots \circ L_{s_{n+1}}(E_\lambda^{n+2}(z)) \\ &= (L_{s_0} \circ L_{s_1} \circ \cdots \circ L_{s_n})(L_{s_{n+1}} \circ E_\lambda(E_\lambda^{n+1}(z))) \\ &= L_{s_0} \circ L_{s_1} \circ \cdots \circ L_{s_n}(E_\lambda^{n+1}(z)) = z. \end{aligned}$$

This shows that $z \in L_{s_0} \circ L_{s_1} \circ \cdots \circ L_{s_n}(B(N))$ for all $n \geq 0$. Therefore, $z = \phi(\overline{s})$ and

$$z = \phi(s_0, s_1, \ldots, s_n, \ldots) \quad \Longrightarrow \quad E_\lambda^n(z) \in R_{s_n}^* \quad \text{for all } n \in \mathbb{N}. \tag{9.7}$$

2. The map ϕ is one-to-one. Let $\overline{s}, \overline{s}^* \in \Sigma_N$ be two different elements of Σ_N such that $\phi(\overline{s}) = \phi(\overline{s}^*) = z$ and let j be the first nonnegative integer such that $s_j \neq s_j^*$. We know that

$$z \in L_{s_0} \circ L_{s_1} \circ \cdots \circ L_{s_n}(B(N)) \cap L_{s_0^*} \circ L_{s_1^*} \circ \cdots \circ L_{s_n^*}(B(N)).$$

Applying E_λ j times, we obtain

$$E_\lambda^j(z) \in L_{s_j}(B(N)) \cap L_{s_j^*}(B(N)),$$

which is a contradiction because the imaginary part of a complex number in $L_{s_j}(B(N))$ belongs to $((2s_j - 1)\pi, (2s_j + 1)\pi)$, while a complex number in $L_{s_j^*}(B(N))$ has it in $((2s_j^* - 1)\pi, (2s_j^* + 1)\pi)$

3. The map ϕ is continuous. Let $\overline{s}^o \in \Sigma_N$, $z_0 = \phi(\overline{s}^o)$, and let $V = \mathbb{D}(z_0, \epsilon) \cap \Lambda_N$ be a neighborhood of z_0 in Λ_N. Recall that $|E_\lambda'(z)| \geq \mu > 1$ for $\mathbb{R}e(z) \geq \ell_\lambda$. Choose a $k \in \mathbb{N}$ such that $\mathrm{d}(B(N))/\mu^{k+1} < \epsilon$ and consider the neighborhood U of $\overline{s}^o$ given by

$$U = \{\overline{s} = (s_i) : s_i = s_i^o, i = 0, 1, \ldots, k\}.$$

By the definition of ϕ,

$$\phi(\overline{s}^o), \phi(\overline{s}) \in L_{s_0} \circ \cdots \circ L_{s_k}(B(N)).$$

For all $z_1, z_2 \in B(N)$, we have

$$|L_{s_0} \circ \cdots \circ L_{s_k}(z_1) - L_{s_0} \circ \cdots \circ L_{s_k}(z_2)| \leq |z_1 - z_2| \sup_{z \in B(N)} |(L_{s_0} \circ \cdots \circ L_{s_k})'(z)|.$$

From this, it is easy to deduce that

$$|\phi(\overline{s}) - \phi(\overline{s}^o)| \leq d(B(N))\frac{1}{\mu^{k+1}} < \epsilon.$$

4. As the set Σ_N is compact, it is clear that ϕ^{-1} is continuous.

5. We need to prove the equality $E_\lambda(z) = \phi \circ \sigma \circ \phi^{-1}(z)$ for all $z \in \Lambda_N$. If $z \in \Lambda_N$ and $(s_0, s_1, s_2, \dots) = \phi^{-1}(z)$, then

$$z \in L_{s_0} \circ L_{s_1} \circ \dots \circ L_{s_n}(B(N)) \quad \text{for all } n \in \mathbb{N}$$

and, therefore, we have

$$E_\lambda(z) \in E_\lambda \circ L_{s_0} \circ L_{s_1} \circ \dots \circ L_{s_n}(B(N)) \quad \text{for all } n \in \mathbb{N}.$$

By (9.3), we conclude

$$E_\lambda(z) \in L_{s_1} \circ \dots \circ L_{s_n}(B(N)) \quad \text{for all } n \in \mathbb{N}.$$

Thus $E_\lambda(z) = \phi(s_1, s_2, \dots) = \phi \circ \sigma(s_0, s_1, s_2, \dots) = \phi \circ \sigma \circ \phi^{-1}(z)$. □

Now we can prove the equality

$$\overline{\bigcup_{n\geq 1} \Lambda_N} = \mathbb{J}(E_\lambda). \tag{9.8}$$

By (9.1), $|E_\lambda'(z)| \geq \mu > 1$ for all $z \in \Lambda_N$. Then Theorem 6.2.3 asserts that Λ_N is contained in the Julia set of E_λ. On the other hand, as $\mathbb{J}(E_\lambda)$ is the closure of the set formed for all repelling periodic points, it suffices to prove that these points belong to $\bigcup_N \Lambda_N$. Thus assume that z_0 is a repelling periodic point for E_λ. Since it belongs to the Julia set, it follows from Corollary 9.1.4 that z_0 must be in the half-plane $\mathbb{R}e(z) \geq \ell_\lambda$. Hence, taking N sufficiently large, the whole repelling cycle is contained in $B(N)$. Obviously, the orbit of z_0 stays in $B(N)$ and, therefore, $z_0 \in \Lambda_N$. This proves the equality (9.8).

In the next section, we go on studying the Julia set of E_λ. Now, using that (E_λ, Λ_N) and (σ, Σ_N) are conjugate, it is very easy to prove the existence of repelling fixed points for E_λ.

We may determine periodic points for σ in a very simple way. If $\overline{s} = (s_0, s_1, \dots, s_n, \dots)$, then $\sigma(\overline{s}) = \overline{s}$ is equivalent to $\overline{s} = (s_0, s_0, \dots, s_0, \dots)$. In consequence, the $2N + 1$ points $z_j = \phi(j, j, \dots, j, \dots)$ with $j = -N, \dots, N$, are fixed points for E_λ. According to (9.7), each z_j belongs to R_j^*. Since we may do this for all $N \in \mathbb{N}$, it follows that E_λ has infinitely many fixed points. As they belong to $\mathbb{J}(E_\lambda)$, these points are necessarily repelling or neutral. Finally, recall that in the previous section we have showed that E_λ has no neutral fixed point.

We may use a similar argument to find periodic points for E_λ. Note that $\sigma^2(\overline{s}) = \overline{s}$ if and only if $\overline{s} = (h, k, h, k, \dots)$, where k and h are integers satisfying $-N \leq k, h \leq N$. Then, for each pair (h,k) with $h \neq k$, $z = \phi(h,k,h,k,\dots)$ is a periodic point for E_λ with period 2 and it is in Λ_N. Again, since there are no neutral periodic points, such points z must be repelling. In this way, we prove that there exist infinitely many repelling periodic points with arbitrary period.

9.4 The Cantor bouquet when $0 < \lambda < 1/e$

Definition 9.4.1. Let $C_N \subset \mathbb{J}(E_\lambda)$ be an invariant set for E_λ. We will say that C_N is a Cantor N-bouquet for E_λ if there exists a homeomorphism

$$h : \Sigma_N \times [1,+\infty) \to C_N,$$

satisfying:

(1) $h(\overline{s},1) = \phi(\overline{s}) = z_{\overline{s}}$.

(2) If $\Pi : \Sigma_N \times [1,+\infty) \to \Sigma_N$ is the natural projection, then

$$\Pi \circ h^{-1} \circ E_\lambda \circ h(\overline{s},t) = \sigma(\overline{s}).$$

(3) $\lim_{t\to+\infty} h(\overline{s},t) = \infty$ for all $\overline{s} \in \Sigma_N$.

(4) If $t \neq 1$, then $\lim_{n\to\infty} E_\lambda^n(h(\overline{s},t)) = \infty$.

To prove the next theorem, we need the following properties of the function $E_{1/e}(t)$ (recall Proposition 9.1.1):

(i) Point 1 is the only real fixed point of $E_{1/e}$.

(ii) $E_{1/e}^n(t) \to +\infty$ for all $t > 1$. Indeed, by induction, it may be proved that the sequence $(E_{1/e}^n(t))_n$ is nondecreasing for each $t > 1$. Since there is only one real fixed point, the sequence cannot be convergent. For the sake of simplicity, we denote $E_{1/e}(t)$ by $E(t)$.

Theorem 9.4.2. *The set $\mathbb{J}(E_\lambda)$ contains a Cantor N-bouquet for all $N \in \mathbb{N}$ and $\lambda \in (0,1/e)$.*

Proof. (I) Given $\lambda \in (0,1/e)$, in order to define $h(\overline{s},t)$, fix $c > 0$ such that $\ell_\lambda < 1 + c < r_\lambda$, and consider the sequence of functions

$$h_k(\overline{s},t) = L_{k,\overline{s}}(E^{k+1}(t) + c),$$

where $L_{k,\overline{s}} = L_{s_0} \circ L_{s_1} \circ \cdots \circ L_{s_k}$. We shall prove that $(h_n(\overline{s},t))$ converges uniformly on $\Sigma_N \times [1,+\infty))$. Note that

$$\begin{aligned}
&|h_{n+1}(\overline{s},t) - h_n(\overline{s},t)| \\
&\quad = |L_{n+1,\overline{s}}(E^{n+2}(t) + c) - L_{n,\overline{s}}(E^{n+1}(t) + c)| \\
&\quad = |L_{n,\overline{s}} \circ L_{s_{n+1}}(E^{n+2}(t) + c) - L_{n,\overline{s}}(E^{n+1}(t) + c)| \\
&\quad = |L_{n\,\overline{s}} \circ L_{s_{n+1}}(e^{E^{n+1}(t)-1} + c) - L_{n,\overline{s}}(E^{n+1}(t) + c)|
\end{aligned}$$

$$= \left|L_{n,\bar{s}}[\log(e^{E^{n+1}(t)-1} + c) - \log\lambda + 2\pi s_{n+1} i] - L_{n,\bar{s}}(E^{n+1}(t) + c)\right|. \tag{9.9}$$

With n and t fixed, applying the Lagrange mean value theorem to $g(x) = \log(e^s + x)$ in $[0, c]$, where $s = E^{n+1}(t) - 1$, we obtain

$$\log(e^{E^{n+1}(t)-1} + c) = E^{n+1}(t) - 1 + cg'(\theta_{n,t}),$$

where $\theta_{n,t} \in (0, c)$. Taking this expression to (9.9), it follows from Claim 1 and (9.3) that

$$\begin{aligned}
&\left|h_{n+1}(\bar{s}, t) - h_n(\bar{s}, t)\right| \\
&\quad = \left|L_{n,\bar{s}}\left(E^{n+1}(t) - 1 - \log\lambda + \frac{c}{e^s + \theta_{n,t}} + 2\pi s_{n+1} i\right) - L_{n,\bar{s}}(E^{n+1}(t) + c)\right| \\
&\quad \le \mu^{-(n+1)}\left|-1 + \log(1/\lambda) + c\left(\frac{1}{e^s + \theta_{n,t}} - 1\right) + 2\pi s_{n+1} i\right| \\
&\quad \le \frac{1 + \log(1/\lambda) + 2\pi N + 2c}{\mu^{n+1}} = \frac{M}{\mu^{n+1}}.
\end{aligned}$$

For $m > n$, this yields

$$\left|h_m(\bar{s}, t) - h_n(\bar{s}, t)\right| \le \left|h_m(\bar{s}, t) - h_{m-1}(\bar{s}, t)\right| + \cdots + \left|h_{n+1}(\bar{s}, t) - h_n(\bar{s}, t)\right| \le \sum_{k=n+1}^{m} \frac{M}{\mu^k}.$$

Denote by h the limit function of h_k as $k \to \infty$.

Later we will need the following estimates of $\mathbb{R}e(h(\bar{s}, t))$ and $\mathbb{I}m(h(\bar{s}, t))$:

$$(2s_0 - 1)\pi \le \mathbb{I}m(h(\bar{s}, t)) \le (2s_0 + 1)\pi, \quad \mathbb{R}e(h(\bar{s}, t)) > t. \tag{9.10}$$

To prove the left inequality, notice that

$$(2s_0 - 1)\pi < \mathbb{I}m(h_n(\bar{s}, t)) < (2s_0 + 1)\pi, \tag{9.11}$$

for all $\bar{s} \in \Sigma_N$, $t \ge 1$ and $n = 0, 1, 2, \ldots$. Then passing to the limit as $n \to \infty$, we obtain the required estimate for $\mathbb{I}m(h(\bar{s}, t))$. For the right inequality, we proceed as follows. For $k = 0$, we have

$$\begin{aligned}
\mathbb{R}e(h_0(\bar{s}, t)) &= \mathbb{R}e(L_{s_0}(E(t) + c)) = \log(E(t) + c) - \log\lambda \\
&= \log(e^{t-1} + c) + \log(1\lambda) \ge t - 1 + \log(1/\lambda) > t,
\end{aligned}$$

since $\log(1/\lambda) > 1$. By induction, assume that the following holds:

$$\mathbb{R}e(h_k(\bar{s}, t)) > t, \quad \text{for all } t \ge 1, \bar{s} \in \Sigma_N, k = 0, 1, \ldots, \tag{9.12}$$

and estimate $\mathbb{R}e(h_{k+1})$ as follows:

$$\begin{aligned}\mathbb{R}e(h_{k+1}(\overline{s},t)) &= \mathbb{R}e(Ls_0 \circ \cdots \circ L_{s_{k+1}}(E^{k+2}(t)+c))\\ &= \mathbb{R}e(L_{s_0}(Ls_1 \circ \cdots \circ L_{s_{k+1}}(E^{k+1}(E(t))+c)))\\ &= \log(|Ls_1 \circ \cdots \circ L_{s_{k+1}}(E^{k+1}(E(t))+c)| - \log(\lambda)\\ &\geq \log(E(t)) - \log(\lambda).\end{aligned}$$

Finally,

$$\mathbb{R}e(h_{k+1}(\overline{s},t)) \geq t - 1 + \log(\lambda) > t.$$

(1) $h(\overline{s},1) = \phi(\overline{s}) = z_{\overline{s}}$. Note that $E^n(1) + c = 1 + c \in B(N)$ for every $n \in \mathbb{N}$. Then

$$h_k(\overline{s},1) = L_{s_0} \circ \cdots \circ L_{s_k}(1+c) \in L_{s_0} \circ \cdots \circ L_{s_k}(B(N)).$$

If $z_{\overline{s}} = \phi(\overline{s})$, $z_{\overline{s}}$ also belongs to the latter set for $k \geq 0$. Therefore,

$$|h_k(\overline{s},1) - z_{\overline{s}}| \leq \operatorname{diam}(L_{s_0} \circ \cdots \circ L_{s_k}(B(N)))$$

and now it suffices to recall that $d(L_{s_0} \circ \cdots \circ L_{s_k}(B(N))) \to 0$ as $k \to \infty$.

Continuity of h. By the uniform convergence, we only need to prove that each h_n is continuous. Given $(\overline{s}^o, t_0)$ and $\varepsilon > 0$, let V be the neighborhood of $\overline{s}^o$ defined by $V = \{(s_n) : s_i = s_i^o \ (\forall i \leq n)\}$. As $L_{n,\overline{s}^o}(E^{n+1}(t)+c)$ is continuous, there exists a $\delta > 0$ such that

$$|L_{n,\overline{s}^o}(E^{n+1}(t)+c) - L_{n,\overline{s}^o}(E^{n+1}(t_0)+c)| < \varepsilon$$

for $t \in U = (t_0 - \delta, t_0 + \delta) \cap [1, +\infty)$. Then, for $(\overline{s}, t) \in V \times U$, we have

$$|h_n(\overline{s},t) - h_n(\overline{s}^o, t_0)| = |L_{n,\overline{s}^o}(E^{n+1}(t)+c) - L_{n,\overline{s}^o}(E^{n+1}(t_0)+c)| < \varepsilon.$$

(2) We prove the equality $E_\lambda(h(\overline{s},t)) = h(\sigma(\overline{s}), E(t))$, which implies condition (2) in Definition 9.4.1. Applying (9.3), we obtain for each $n \in \mathbb{N}$,

$$\begin{aligned}E_\lambda(h_n(\overline{s},t)) &= E_\lambda(L_{s_0} \circ \cdots \circ L_{s_n}(E^{n+1}(t)+c)) = L_{\sigma(\overline{s}),n-1}(E^{n+1}(t)+c)\\ &= L_{\sigma(\overline{s}),n-1}(E^n(E(t))+c) = h_{n-1}(\sigma(\overline{s}), E(t)).\end{aligned}$$

Passing to the limit as $n \to \infty$, we deduce the required result.

Injectivity of h. Suppose that $h(\overline{s},t) = h(\overline{s}^*, t^*) = z$ but $(\overline{s},t) \neq (\overline{s}^*, t^*)$. If we show that $s_0 = s_0^*$, then it is easy to obtain the equality $\overline{s} = \overline{s}^*$. Indeed, it follows from (2) that

$$h(\sigma(\overline{s}), E(t)) = h(\sigma(\overline{s}^*), E(t^*)),$$

and this yields $s_1 = s_1^*$. Proceeding in this way, we get the equality $\overline{s} = \overline{s}^*$. So, let us prove that $s_0 = s_0^*$. By (9.11),

$$(2s_0 - 1)\pi \leq \mathbb{I}m(z) \leq (2s_0 + 1)\pi \quad \text{and} \quad (2s_0^* - 1)\pi < \mathbb{I}m(z) < (2s_0^* + 1)\pi,$$

but this is only possible if either $s_0 = s_0^*$ or they are consecutive integers and $\mathbb{I}m(z) = 2\pi s + 1$, with s being an integer. In the latter case, we have

$$E_\lambda(z) = \lambda e^{\mathbb{R}e(z)} e^{(2s+1)\pi i} = -\lambda e^{\mathbb{R}e(z)} < 0$$

which contradicts property (2).

Continuity of h^{-1}. Put $\Gamma_N = h(\Sigma_N \times [1, \infty))$. First, notice that, given $T > 1$, the restriction map h_T^{-1} of h^{-1} to $h(\Sigma_N \times [1, T])$ is continuous since $\Sigma_N \times [1, T]$ is compact. Now suppose that $z_n \to z_0$, where $z_n, z_0 \in \Gamma_N$. Then $z_n = h(\overline{s}_n, t_n)$ and $z_0 = h(\overline{s}_0, t_0)$. By (9.10), $\mathbb{R}e(h(\overline{s}_n, t_n)) > t_n$ for all $n \in \mathbb{N}$. As $(z_n = h(\overline{s}_n, t_n))$ is bounded, this implies that there exists $T > 1$ such that $t_n < T$ for all n, and we are in the previous case.

(3) $\lim_{t\to+\infty} h(\overline{s}, t) = \infty$. We have seen in (1) that $\mathbb{R}e(h(\overline{s}, t)) \geq t$ for every $\overline{s} \in \Sigma$ and $t \geq 1$. Therefore, $\lim_{t\to+\infty} \mathbb{R}e(h(\overline{s}, t)) = +\infty$.

(4) *If $t \neq 1$, then* $\lim_{n\to\infty} \mathbb{R}e(E_\lambda^n(h(\overline{s}, t))) = +\infty$. It follows from (9.3) that

$$E_\lambda^n(h_k(\overline{s}, t)) = L_{s_n} \circ \cdots \circ L_{s_k}(E^{k+1}(t) + c) = L_{s_n} \circ \cdots \circ L_{s_k}(E^{k-n+1}(E^n(t)) + c)$$

for $k, n \in \mathbb{N}$ with $n \leq k$. Applying (9.12), we deduce that

$$\mathbb{R}e(E_\lambda^n(h_k(\overline{s}, t))) = \mathbb{R}e(L_{s_n} \circ \cdots \circ L_{s_k}(E^{k-n+1}(E^n(t)) + c)) > E^n(t).$$

Finally, passing to the limit as $k \to \infty$, with $n, t, \overline{s}$ fixed, we obtain

$$\mathbb{R}e(E_\lambda^n(h(\overline{s}, t))) \geq E^n(t).$$

As $t > 1$, the proof concludes. □

9.5 Hairs in the dynamical plane

Given $\overline{s} \in \Sigma_N$, consider the curve $h_{\overline{s}} : [1, +\infty) \to \mathbb{C}$ defined by $h_{\overline{s}}(t) = h(\overline{s}, t)$. We may interpret it as a "hair" that is born at $z_{\overline{s}} = h_{\overline{s}}(1) = h(\overline{s}, 1) \in \Lambda_N$ with the following properties:

(H1) $h_{\overline{s}}(1) = z_{\overline{s}}$.

(H2) For each $t \geq 1$, the itinerary of $h_{\overline{s}}(t)$ under E_λ is $\overline{s}$. Indeed, by property (2) in Definition 9.4.1, we have $E_\lambda^n(h_{\overline{s}}(t)) = h(\sigma^n(\overline{s}), E^n(t))$. So, (9.10) tells us that

$$(2s_n - 1)\pi \leq \mathbb{I}m(E_\lambda^n(h_{\overline{s}}(t))) \leq (2s_n + 1)\pi,$$

which implies that $E_\lambda^n(h_{\overline{s}}(t)) \in R^*(s_n)$.

(H3) If $t > 1$, then $\lim_{n\to\infty} \mathbb{R}e(E_\lambda^n(h_{\overline{s}}(t))) = +\infty$.

(H4) $\lim_{t\to\infty} \mathbb{R}e(h_{\overline{s}}(t)) = +\infty$.

(H5) The curve $h_{\overline{s}}$ lies in the right half-plane and is contained in the strip $S(s_0)$. Again this follows from (9.10).

Roughly speaking, the hair attached to $z_{\overline{s}}$ is a continuous curve stretching from $z_{\overline{s}}$ to ∞ in the right half-plane. Every point z on the hair shares the same itinerary as $z_{\overline{s}}$ and has an orbit which tends to ∞ in the right half-plane. However, the orbit of the endpoint $z_{\overline{s}}$ is bounded.

By property (2) in Definition 9.4.1, E_λ permutes the hairs as σ permutes the sequences in Σ_N. Indeed, it holds

$$E_\lambda(h(\overline{s}, t)) = h(\sigma(\overline{s}), E(t)).$$

It is easy to prove that, for each $\overline{s} \in \Sigma_N$, there exists a unique hair attached to $z_{\overline{s}}$. In fact, suppose that $h(t)$ is another continuous curve stretching from $z_{\overline{s}}$ to ∞ in the right half-plane, satisfying properties (H1)–(H5), and denote by U the open nonempty set limited by the two hairs. Then $\mathbb{F}(E_\lambda) = (\mathbb{F}(E_\lambda) \cap U) \cup (\mathbb{F}(E_\lambda) \setminus (\overline{U}))$, which implies that $\mathbb{F}(E_\lambda)$ is not connected and this contradicts the fact that $\mathbb{F}(E_\lambda) = \Omega_0(E_\lambda, q_\lambda)$.

The set Γ_N is called a Cantor N-bouquet. Note that condition (2) in the definition asserts that each set Γ_N is invariant for E_λ. The set $\Gamma = \overline{\bigcup_N \Gamma_N}$ is called the Cantor bouquet of E_λ.

By property (H3), the points $h(\overline{s}, t)$ have divergent orbits for $t > 1$. According to Theorem 8.3.4, these points belong to $\mathbb{J}(E_\lambda)$ because E_λ only has one singular value.

The remaining points $\{h(\overline{s}, 1) : \overline{s} \in \Sigma_N\} = \Lambda_N$ are in the Julia set, too, as we have already seen, but these points have bounded orbit. The set Λ_N is called the crown of the bouquet.

Finally, it is easy to prove the equality $\mathbb{J}(E_\lambda) = \Gamma$. Indeed, we already know the inclusion $\Gamma_N \subset \mathbb{J}(E_\lambda)$, and the reverse inclusion follows from (9.8).

9.6 The explosion of the exponential

We know that the origin is the unique singular value of E_λ, therefore, as we have seen in the previous chapter, the orbit of this omitted point is the crucial factor which governs its dynamics.

Proposition 9.6.1. *If the orbit of $z_0 = 0$ converges to ∞ or is preperiodic, then $\mathbb{J}(E_\lambda) = \mathbb{C}$.*

Proof. As E_λ has no superattracting cycles, Theorem 8.3.3 tells us that $\mathbb{J}(E_\lambda) = \mathbb{C}$. □

Corollary 9.6.2. *If $\lambda > 1/e$ or $\lambda = k\pi i$ with $k \in \mathbb{Z}$, then $\mathbb{J}(E_\lambda) = \mathbb{C}$.*

Proof. If $\lambda > 0$, by induction, it is easy to prove that the orbit of the origin is increasing. Then $(E_\lambda^n(0))_n$ necessarily converges to a fixed point of E_λ or ∞. But, for $\lambda > 1/e$, E_λ has no real fixed points. Thus the orbit of $z_0 = 0$ converges to ∞.

If $\lambda = k\pi i$, then the orbit of the origin is the sequence

$$0,\ k\pi i,\ (-1)^k k\pi i,\ (-1)^k k\pi i, \ldots$$

and, therefore, preperiodic. □

The value $\lambda_0 = 1/e$ has the property that $\mathbb{J}(E_\lambda) = \mathbb{C}$ for $\lambda > \lambda_0$ and $\mathbb{J}(E_\lambda)$ is nowhere dense for $0 < \lambda < \lambda_0$. That is the reason why $\lambda_0 = 1/e$ is called an *explosive value* for the family E_λ.

In Figure 9.2, the color of a point depends on the escaping speed of the orbit to the infinity under the function E_λ for $\lambda = 0.3686 < 1/e$. In this case, the Julia set is the complex plane and the chaotic behavior of orbits results in an aesthetic image. In particular, if $\lambda = 0.3578 < 1/e$, Figure 9.3 shows the filled Julia set of E_λ (black).

Figure 9.2: The set $\mathbb{J}(f) = \mathbb{C}$ for $f(z) = 0.3686 \exp(z)$.

Figure 9.3: Filled Julia set (black) of $f(z) = 0.3678 \exp(z)$.

Proposition 9.6.3. *There exists a sequence $\lambda_n \to 1$ such that $\mathbb{J}(E_{\lambda_n}) = \mathbb{C}$ for every $n \in \mathbb{N}$.*

Proof. Let $g(\lambda) = \lambda e^{\lambda}$ for $\lambda \in \mathbb{C}$. Notice that $g^n(\lambda) = E_{\lambda}^{n+1}(0)$ for all n. By Example 8.1.2, g belongs to the class $\mathcal{S}$ and, in consequence, every point whose orbit under g converges to ∞ must belong to the Julia set of g. Since $g^n(1) \to \infty$, it follows that (g^n) is not normal in any neighborhood of 1. According to Montel's theorem, for each $n \in \mathbb{N}$, there exist a value λ_n arbitrarily close to 1 and $k_n \in \mathbb{N}$ such that $g^{k_n}(\lambda_n) = 2\pi m i$ for some $m \in \mathbb{Z}$. Then $E_{\lambda_n}^{k_n+1}(0) = g^{k_n}(\lambda_n) = 2\pi m i$, which implies that $E_{\lambda_n}^{k_n+2}(0) = \lambda_n = E_{\lambda_n}(0)$. So, 0 is preperiodic for E_{λ_n} and the above proposition asserts that $\mathbb{J}(E_{\lambda_n}) = \mathbb{C}$. □

9.7 Exercises

1. Using only the results of Section 9.1, prove that the attractive basin of q_{λ} is dense in $\mathbb{C}$.
   Hint: Apply Block's theorem: If f is analytic on $\overline{\mathbb{D}}$ and verifies $f'(0) = 1$, then there exists an absolute constant $L > 0$ such that $f(\overline{\mathbb{D}})$ contains a disc with radius L [45, pp. 385–388].
2. Let $f(z) = e^z$ and $S = \{z \in \mathbb{C} : |\mathbb{I}m(z)| \le \pi/3\}$. Prove that

$$\mathbb{R}e(f(z)) \ge \mathbb{R}e(z) + (1 - \log 2),$$

   for all $z \in S$, and deduce that if $f^n(z) \in S$ for all $n \in \mathbb{N}$, then the orbit of z is divergent. In particular, this is the case if z is real.
3. (Misiurewicz) Under the conditions of the previous exercise, let z_0 be a nonreal complex number. Show that there exists an $m \in \mathbb{N}$ such that $f^m(z_0) \notin S$.
4. Show that $f(z) = e^z$ has a repelling fixed point in each strip $S_k = \{z = z + iy : (2k-1)\pi < \mathbb{I}m(z) < (2k+1)\pi\}$ if $k \neq 0$.
5. Consider the function $f(z) = e^{z-1}$. Prove the following statements:
   (i) The unit disc $\mathbb{D}$ is invariant under f.
   (ii) f maps the half-plane $\mathbb{R}e(z) < 1$ into the disc $\mathbb{D}$.
   (iii) The half-plane $\mathbb{R}e(z) < 1$ is contained in $\mathbb{F}(f)$.
   The following three exercises deal with the Julia set of the entire function $S_{\lambda}(z) = \lambda \sin z$.
6. Assume $0 < \lambda < 1$ and prove the following statements:
   (i) The imaginary axis is invariant under S_{λ}.
   (ii) The function S_{λ} has 3 fixed points on the imaginary axis: the origin (attracting) and the points ip_+ and ip_-, for $p_+ > 0$ and $p_- = -p_+$ (repelling).
   (iii) $S_{\lambda}^n(iy) \to \infty$ for every $y > p_+$ and then $\{yi : y \ge p_+\} \subset \mathbb{J}(S_{\lambda})$.
7. Given $\lambda \in (0,1)$, $N \in \mathbb{N}$, and $\varepsilon > 0$, consider the strips $W_j = \{z = x + yi : (2j-1)\pi < x < (2j+1)\pi, y > \varepsilon\}$ for $j = 0, \pm 1, \dots, \pm N$. Put $B(N) = \bigcup_{0 \le |j| \le N} W_j$ and $\Delta_N = \{z \in B(N) : S_{\lambda}^n(z) \in B(N)\ (\forall n \in \mathbb{N})\}$. Prove the following statements:
   (i) The set Δ_N is nonempty and invariant.

(ii) For every $z \in \Delta_N$, there is a sequence (s_n) such that $S_\lambda^n(z) \in W_{s_n}$ for all n (this (s_n) is called the itinerary of z).

(iii) $\varepsilon > 0$ may be chosen so that $\Delta_N \subset \mathbb{J}(S_\lambda)$.

8. Suppose that $|\lambda| \geq 1$ and prove that the imaginary axis is contained in $\mathbb{J}(S_\lambda)$.
9. Let $f(z) = \lambda e^{\sin z}$ with $|\lambda| \leq 1/17$. Show that:
   (i) The function f has at most one asymptotic value.
   (ii) The set $\mathbb{D}$ is invariant under f.
   (iii) $\mathbb{F}(f) = \Omega(f, 0)$.
10. Describe the dynamics of $f(z) = (1/e)e^z$.

Bibliography remarks

Section 9.2 is inspired by [28, Section 3.5].

Section 9.3 is based on [27, pp. 181–206] and [63, Section 3.2]. The results related to the Cantor bouquet are only sketched in most books.

In Section 9.6, we follow [47, Section 5.4].

10 The parameter plane

10.1 Introduction

The study of parametric families forms a major part of complex dynamics. Given a family of entire functions $f_c(z)$, we can do a partition of the parameter plane into disjoint regions Ω_k in such a way that when c moves through each Ω_k, the dynamics of the iterations of f_c display essentially the same features, while, in contrast, when c passes from one Ω_k to another, a significant change in the dynamics takes place.

The quadratic family $Q_c(z) = z^2 + c$ is the first and most important example. We can create computer generated graphics in the parameter plane (the c-plane) by assigning to each c a color which depends on the dynamics of the function Q_c. The first graphs of this type were obtained by Brooks and Matelski (1978, [19]). About the same time, Mandelbrot created graphics for the family $P_\lambda(z) = \lambda z(1-z)$. Therefore, in order to prevent confusion, we translate all of Mandelbrot's definitions to the family $Q_c(z) = z^2 + c$. He introduced the set $\mathbb{M}$ formed by all c such that the orbit $(Q_c^n(0))_n$ is bounded. The images created by Mandelbrot seem to show a number of isolated subsets of $\mathbb{M}$. Thus he conjectured that $\mathbb{M}$ had many distinct connected components. The first real mathematical breakthrough came with Douady and Hubbard (1982) when they introduced the name Mandelbrot set for the set $\mathbb{M}$ and provided a solid foundation for its study. Among other results, they proved that $\mathbb{M}$ and its complement are connected. The study of the parameter plane for cubic polynomial families began five years later with the work of Branner and Hubbard, who considered the cubic family $P(z) = z^3 - 3a^2z + b$ [18]. Polynomial families of a higher degree have been studied by Lavaurs [57].

10.2 The parameter plane

Given a family $\{f_c : c \in \mathbb{C}\}$ of entire functions, if z_0 is the only singular value of f_c for all $c \in \mathbb{C}$, the set

$$\mathbb{P} = \{c \in \mathbb{C} : (f_c^n(z_0))_n \text{ is bounded}\}$$

is called the *parameter plane* of the family. This set is a catalogue of the dynamics of the family. We can consider the following mutually disjoint subsets in $\mathbb{P}$. For each $k \in \mathbb{N}$, denote by $C(k)$ the set of all $c \in \mathbb{C}$ such that the orbit $(f_c^n(z_0))$ is attracted by an attracting cycle for f_c with period k. Obviously, every $C(k)$ is contained in $\mathbb{P}$. Each component of $C(k)$ is called a *hyperbolic component* (in Figure 10.1, we see the hyperbolic components of the quadratic family).

https://doi.org/10.1515/9783111689685-010

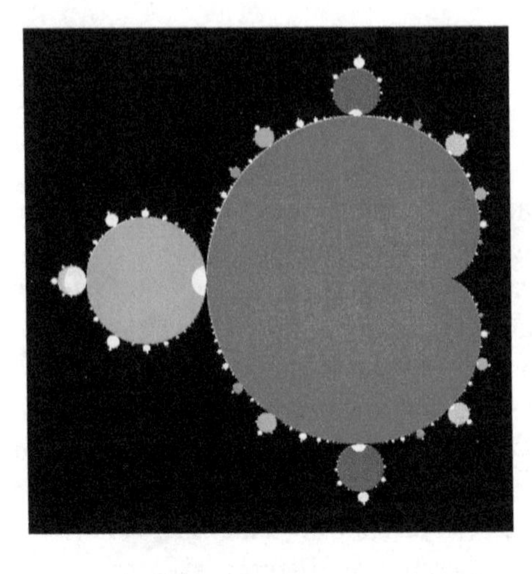

Figure 10.1: Mandelbrot set showing the sets $C(k)$ for $k = 1, \dots, 6$ (red, green, blue, pink, ...). The yellow region is formed by all c belonging to the Mandelbrot set such that the used algorithm cannot decide if the origin is attracted by an attracting orbit with period less or equal to 6.

Next we will prove some general properties of the sets $C(k)$. We assume the family $\{f_c : c \in \mathbb{C}\}$ satisfies the following conditions:
(C1) For each $k \in \mathbb{N}$, the map $(c,z) \in \mathbb{C} \times \mathbb{C} \to (f_c^k)'(z) \in \mathbb{C}$ is continuous.
(C2) For every $z \in \mathbb{C}$ and $k \in \mathbb{N}$, the map $c \in \mathbb{C} \to f_c^k(z) \in \mathbb{C}$ is entire.
(C3) The point z_0 is the only singular value of f_c for every $c \in \mathbb{C}$.

Proposition 10.2.1. *Under the above conditions, each $C(k)$ is open.*

Proof. By contradiction, suppose $C(k)$ is not open. Then there exist $c_0 \in C(k)$, $c_n \notin C(k)$ such that $c_n \to c_0$. As $c_0 \in C(k)$, f_{c_0} has a k-periodic attracting point z^*. Hence

$$f_{c_0}^k(z^*) = z^* \quad \text{and} \quad |(f_{c_0}^k)'(z^*)| < 1. \tag{10.1}$$

Since $(f_c^k)'(z)$ is continuous in $\mathbb{C} \times \mathbb{C}$, there is a neighborhood $U = \mathbb{D}(c_0) \times \overline{\mathbb{D}}_r(z^*)$ of (c_0, z^*) such that

$$|(f_c^k)'(z)| < 1 \quad \text{for all } (c,z) \in U. \tag{10.2}$$

Choose an $n_0 \in \mathbb{N}$ such that $c_n \in \mathbb{D}(c_0)$ for $n \geq n_0$. Now, for each $n \geq n_0$, put $g_n(z) = f_{c_n}^k(z) - z$. We will show that there exists an $m_0 \geq n_0$ such that $(g_n)_{n \geq m_0}$ is uniformly bounded in $\mathbb{D}_r(z^*)$ and, therefore, normal in $\mathbb{D}_r(z^*)$. In fact, using (10.2), we obtain

$$|g_n(z) - g_n(z^*)| \leq |z - z^*| \sup\{|g_n'(w)| : w \in \mathbb{D}_r(z^*)\} \leq 2r,$$

for $n \geq n_0$ and $z \in \mathbb{D}_r(z^*)$.

On the other hand, (10.1) gives $\lim_{n\to\infty} g_n(z^*) = f_{c_0}^k(z^*) - z^* = 0$. Then there exists an $m_0 \geq n_0$ such that $|g_n(z^*)| \leq 1$ for $n \geq m_0$. Now it is evident that

$$|g_n(z)| \leq 2r + 1 \quad \text{for all } n \geq m_0 \text{ and } z \in \mathbb{D}_r(z^*).$$

Since $(g_n)_{n\geq m_0}$ is normal on $\mathbb{D}_r(z^*)$, there exists a subsequence $(g_{n_p})_p$ uniformly convergent on compact subsets of $\mathbb{D}_r(z^*)$ to $g(z) = f_{c_0}^k(z) - z$. Finally, notice that $g(z^*) = 0$ but $g_n(z) \neq 0$ in $\mathbb{D}_r(z^*)$, which contradicts Hurwitz's theorem. □

If C is a connected component of $C(k)$, for each $c \in C, f_c$ has an attracting k-periodic orbit. As z_0 is the only singular value of f_c, z_0 belongs to the attractive basin. Then $(f_c^{kn}(z_0))_n$ converges to a k-periodic point $z^*(c)$. We will prove that the function $z^*(c)$ is analytic in C.

Proposition 10.2.2. *Under the above conditions, let C be a connected component of $C(k)$. The following statements hold:*
(i) *The function $c \in C \to z^*(c) \in \mathbb{C}$ is analytic.*
(ii) *The function $\chi(c) = (f_c^k)'(z^*(c))$ is analytic in C.*

Proof. (i) Consider the sequence of functions (G_n) defined by $G_n(c) = f_c^n(z_0)$. According to (C2), the G_n's are entire functions and we know that $\lim_{n\to\infty} G_{kn}(c) = z^*(c)$ for each $c \in C$. We will prove that $(G_{kn})_n$ is normal in C showing that (G_n) is uniformly bounded in a neighborhood of every $c \in C$. For this, for a fixed $c_0 \in C$, let $\mu > 0$ be such that $|(f_{c_0}^k)'(z^*(c_0))| < \mu < 1$. By continuity, there exists a neighborhood $V(c_0) \times \mathbb{D}_r(z^*(c_0))$ of $(c_0, z^*(c_0))$ such that

$$|(f_c^k)'(z)| < \mu < 1 \quad \text{for all } c \in V(c_0),\ z \in \mathbb{D}_r(z^*(c_0)), \tag{10.3}$$

where we take $V(c_0)$ satisfying $V(c_0) \subset C$. Now choose $V(c_0)$ sufficiently small such that

$$|f_c^k(z^*(c_0)) - f_{c_0}^k(z^*(c_0))| < r(1-\mu) \quad \text{for all } c \in V(c_0).$$

Then, for $z \in \mathbb{D}_r(z^*(c_0))$ and $c \in V(c_0)$, we have

$$\begin{aligned} &|f_c^k(z) - f_{c_0}^k(z^*(c_0))| \\ &\quad \leq |f_c^k(z) - f_c^k(z^*(c_0))| + |f_c^k(z^*(c_0)) - f_{c_0}^k(z^*(c_0))| \\ &\quad \leq \sup\{|(f_c^k)'(w)| : w \in \mathbb{D}_r(z^*(c_0))\}|z - z^*(c_0)| + r(1-\mu) < r. \end{aligned}$$

This proves that, for each $c \in V(c_0)$,

$$f_c^k(\mathbb{D}_r(z^*(c_0))) \subset \mathbb{D}_r(z^*(c_0)).$$

Since $f_{c_0}^{kn}(z_0) \to z^*(c_0)$ as $n \to \infty$, it follows that there exists an $m \in \mathbb{N}$ satisfying $f_{c_0}^{km}(0) \in \mathbb{D}_r(z^*(c_0))$. Finally, by the continuity of $f_c^{mk}(z_0)$ in C, there is a neighborhood $W(c_0)$ of c_0 such that

$$f_c^{mk}(z_0) \in \mathbb{D}_r(z^*(c_0)) \quad \text{for all } c \in W(c_0).$$

If $V_0 = V(c_0) \cap W(c_0)$, we have

$$f_c^{mk}(z_0) \in \mathbb{D}_r(z^*(c_0)) \quad \text{and} \quad f_c^k(\mathbb{D}_r(z^*(c_0))) \subset \mathbb{D}_r(z^*(c_0)),$$

for $c \in V_0$. From the above conditions, it follows that $G_{kn}(c) \in D_r(z^*(c_0))$ for $n \geq m$ and $c \in V_0$. Therefore, $(G_{nk})_n$ is locally uniformly bounded in C. With the normality in hand, the proof concludes.

(ii) Note that $\chi(c)$ is the composition of the functions $z^*(c)$ and $(f_c^k)'(z)$. □

Proposition 10.2.3. *Every connected component of $C(k)$ is simply connected.*

Proof. Let C be a connected component of $C(k)$ and let γ be a Jordan curve in C. If K is the compact region delimited by γ, we will prove that $K \subset C$. Take an open neighborhood U of γ such that $\overline{U} \subset C$. As in the proof of the above proposition, let $G_n(c) = f_c^n(z_0)$ and recall that $(G_{kn})_n$ is normal in C. By the Cauchy integral formula, we have

$$G_{kn}(c) = \frac{1}{2\pi i} \oint_\gamma \frac{G_{kn}(\zeta)\, d\zeta}{\zeta - c} \quad \text{for } n \in \mathbb{N} \text{ and } c \in \mathring{K} \setminus \overline{U}. \tag{10.4}$$

Since $(G_{kn})_n$ is normal in C and $\lim_{n\to\infty} G_{kn}(c) = z^*(c)$ for every $c \in C$, it follows that this convergence is uniform on compact subsets of C. This, together with (10.4), implies that $G_{kn}(c)$ converges uniformly on $\mathring{K} \setminus \overline{U}$ to an analytic function that we denote by $G(c)$. This function is defined in $C \cup K$ and satisfies $G(c) = z^*(c)$ for $c \in C$. If we put $\chi(c) = (f_c^k)'(G(c)), \chi(c)$ is also defined in $C \cup K$ and $\chi(c)$ is the multiplier of $z^*(c)$ for $c \in C$. As $|\chi(c)| < 1$ for $c \in \gamma$, we deduce from the maximum modulus principle that $|\chi(c)| < 1$ in K. Finally, note that $f_c^k(G(c)) = G(c)$ in U and, therefore, the identity principle tells us that the above equality is valid for $c \in K$. Then we have proved that, for all $c \in K$, $G(c)$ is an attracting periodic point for f_c and to conclude the proof we only need to show that, for all $c \in \mathring{K}$, k is the period of $G(c)$. Otherwise, there exists a $c_0 \in \mathring{K}$ such that $G(c_0)$ is m-periodic for f_{c_0} with $m < k$. As $C(m)$ is open, there is a neighborhood $V(c_0)$ of c_0 such that $V(c_0) \subset C_m \cap \mathring{K}$. The equality $f^m(G(c)) = G(c)$ holds for $c \in V(c_0)$ and, by the identity principle, it is also valid in $\mathring{K}$. This is not possible in $\mathring{K} \cap U$. □

In Figures C.15–C.18, we show the parameter plane of several families $f_c(z)$.

10.3 The Mandelbrot set

The Mandelbrot set is one of the most spectacular and intricate fractal sets. It was discovered by B. Mandelbrot at the end of the 1970s. It is the parameter plane of the quadratic family $Q_c(z) = z^2 + c$. As the origin is the only critical point of Q_c, it is defined by (recall Theorem 8.2.1)

$$\mathbb{M} = \{c \in \mathbb{C} : (Q_c^n(0))_n \text{ is bounded}\}.$$

The next theorem places the Mandelbrot set in the disc $\overline{\mathbb{D}}_2(0)$ and gives us a useful stop test, in case we want to obtain an approximate computer generated graph of $\mathbb{M}$.

Theorem 10.3.1. *One has* $\mathbb{M} = \{c \in \mathbb{C} : |Q_c^n(0)| \le 2\ (\forall n \in \mathbb{N})\}$.

Proof. The inclusion

$$\{c \in \mathbb{C} : |Q_c^n(0)| \le 2\ (\forall n \in \mathbb{N})\} \subset \mathbb{M}$$

is evident. For the converse inclusion, first we show that $\mathbb{M} \subset \overline{\mathbb{D}}_2(0)$. Take $c \in \mathbb{M}$ and suppose $|c| > 2$. Then the sequence $(Q^n(c))_n$ is bounded and, therefore, $c \in \mathbb{B}(Q_c)$. It follows from Theorem 6.6.2 that $|Q_c^n(c)| \le R(c)$ for $n = 0, 1, 2, \ldots$, where $R(c) = \max\{2, \sqrt{2|c|}\}$. In particular, $|c| \le R(c)$. As $|c| > 2$, we have $R(c) = \sqrt{2|c|}$ and this yields $|c| \le \sqrt{2|c|}$ or, equivalently, $|c| \le 2$. Thus we have reached a contradiction.

Finally, assume $c \in \mathbb{M}$. From the previous step, we know that $|c| \le 2$. By contradiction, if there is an $m \in \mathbb{N}$ such that $|Q_c^m(0)| > 2$, then, as $R(c) = 2$, Theorem 6.6.2 applied to $w = Q_c^m(0)$ yields

$$\infty = \lim_{n\to\infty} Q_c^n(w) = \lim_{n\to\infty} Q_c^{m+n}(0).$$

This is not possible since $c \in \mathbb{M}$. □

The next proposition shows that $-2 \in \mathbb{M}$, then the above result is the best of its type. The proof is Exercise 2.

Proposition 10.3.2. *The intersection of* $\mathbb{M}$ *with the real axis is the interval* $[-2, 1/4]$.

Theorem 10.3.3. *The set* $\mathbb{M}$ *is compact and its complement in* $\hat{\mathbb{C}}$ *is connected.*

Proof. Inductively, we define the sequence of polynomials $(P_n(c))_n$ as follows: $P_1(c) = c$ and $P_{n+1}(c) = P_n^2(c) + c$. Note that $Q_c^{n+1}(0) = Q_c^n(c) = P_n(c)$. By Theorem 10.3.1, we have

$$\mathbb{M} = \bigcap_{n=1}^{\infty} P_n^{-1}(\overline{D}_2(0)),$$

which proves that $\mathbb{M}$ is closed and, therefore, compact.

If we set $P_n(\infty) = \infty$, the P_n's are continuous functions defined from $\hat{\mathbb{C}}$ onto itself. Taking complement, we obtain

$$\hat{\mathbb{C}} \setminus \mathbb{M} = \bigcup_{n=1}^{\infty} P_n^{-1}(V),$$

where $V = \{\infty\} \cap \{c \in \mathbb{C} : |c| > 2\}$. The sequence $(P_n^{-1}(V))$ is nondecreasing and $V \subset P_n^{-1}(V)$, for all $n \in \mathbb{N}$ since $|Q_c(c)| > 2$ for every $|c| > 2$. By a similar argument as in the proof of Theorem 6.5.5, we may prove that $P_n^{-1}(V)$ is connected. In consequence, the complement of $\mathbb{M}$ is connected, too. □

Now our aim is to find the relation between the Mandelbrot set and the dynamics of the quadratic family. We will show that the Mandelbrot set is like a catalogue of the dynamics of Q_c.

Theorem 10.3.4. *For every $|\lambda| \le 1$, there is a unique constant $c(\lambda)$ such that $Q_{c(\lambda)}$ has a fixed point z_λ^* satisfying $Q'_{c(z_\lambda^*)}(z_\lambda^*) = \lambda$. If $|\lambda| < 1$, then $c(\lambda)$ belongs to the interior of the cardioid C given by*

$$C : r_c = (1/4)(5 - 4\cos\theta)^{1/2} \quad (\theta \in [0, 2\pi]),$$

and z_λ^* is an attracting fixed point of $Q_{c(\lambda)}$. If $|\lambda| = 1$, then $c(\lambda)$ is in C and z_λ^* is a neutral fixed point for $Q_{c(\lambda)}$.

Proof. First, we obtain the fixed points of Q_c in terms of c. The equation $z^2 + c = z$ has two solutions, $(1 \pm \sqrt{1-4c})/2$. As we want λ to be the multiplier of the fixed point, $\lambda = 1 + \sqrt{1-4c}$. That is,

$$c(\lambda) = \lambda/2 - \lambda^2/4. \tag{10.5}$$

If $\overline{\mathbb{D}}$ denotes the closed unit disc $\{\lambda : |\lambda| \le 1\}$, let

$$A = \{c = \rho e^{i\theta}/2 - \rho^2 e^{2i\theta}/4 : 0 \le \rho \le 1, 0 \le \theta \le 2\pi\}$$

and consider the map defined by

$$h : \lambda \in \overline{\mathbb{D}} \to c(\lambda) = \lambda/2 - \lambda^2/4 \in A.$$

Reasoning by contradiction, it is easy to prove that h is bijective.

If we express λ in the form $\lambda = \rho e^{i\theta}$, for $\rho \in [0,1]$ and $\theta \in [0, 2\pi]$, then (10.5) may be rewritten as

$$c = (1/4)(2\rho e^{i\theta} - \rho^2 e^{2\theta i}). \tag{10.6}$$

For each $\rho \in (0,1]$, (10.6) is the equation of a cardioid, and we obtain the largest one when $\rho = 1$. The equation in polar coordinates is

$$r_c = (1/4)(4\rho^2 + \rho^4 - 4\rho^3\cos\theta)^{1/2}, \quad \rho \in (0,1],\ \theta \in [0, 2\pi],$$

where r_c denotes the modulus of c. For $\rho = 1$, we have the cardioid C, namely

$$r_c = (1/4)(5 - 4\cos\theta)^{1/2}. \tag{10.7}$$

The cardioid C intersects the real axis in two points, $\theta = 0, \pi$. Thus these points are $z = -3/4, 1/4$.

As $1-4c(\lambda) = (\lambda-1)^2$, if we put $z_\lambda^* = (1+\sqrt{1-4c(\lambda)})/2$, then $z_\lambda^* = \lambda/2$ and $Q'_{c(\lambda)}(z_\lambda^*) = \lambda$. Therefore, if $|\lambda| < 1$, z_λ^* is an attracting fixed point for $Q_{c(\lambda)}$ and $c(\lambda)$ is interior to the cardioid C. If $|\lambda| = 1$, $c(\lambda)$ belongs to the cardioid C and z_λ^* is a neutral fixed point for $Q_{c(\lambda)}$.

On the other hand, the fixed point $z_\lambda^{**} = (1-\sqrt{1-4c(\lambda)})/2 = 1-\lambda/2$ is repelling for $\lambda \neq 1$ because $Q'_{c(\lambda)}(z_\lambda^{**}) = 2-\lambda$ (note, if $\lambda = 1$, then $z_\lambda^* = z_\lambda^{**}$). □

The above theorem shows that the hyperbolic component $C(1)$ is the interior of the cardioid C. Next, we see that the set $C(2)$ is the disc $\mathbb{D}_{1/4}(-1)$. This is Exercise 9.

Theorem 10.3.5. *The set formed by all c such that Q_c has an attracting 2-cycle is the disc $\mathbb{D}_{1/4}(-1)$.*

The dynamics of Q_c is also well known when c varies in other subsets of $\mathbb{M}$. In Figure 10.1, we see an image of the Mandelbrot set showing the sets $C(k)$ with different colors. Along the main cardioid, M presents a series of bulbs of various sizes. We point out the two following cases:

- If c is in one of the two greater bulbs placed in the highest and lowest part of the cardioid (blue bulbs in Figure 10.1), then Q_c has an attracting 3-periodic orbit.
- If c belongs to one of the two biggest bulbs placed on the right part of the cardioid (pink bulbs in the same figure), Q_c has an attracting 4-periodic orbit.

Now we are interested in the values of the parameter c for which Q_c has a superattracting periodic orbit. As we have mentioned earlier, Q_c has only one attracting periodic orbit because the origin is the unique critical point. Then, if $\{z_0, z_1, \dots, z_{m-1}\}$ is a superattracting m-cycle for Q_c, as $(Q_c^m)'(z_0) = \prod_0^{m-1}(Q_c)'(z_i) = 0$, we deduce that the origin is one of the points in the cycle. In other words, Q_c has a superattracting periodic orbit if and only if the origin is a periodic point for Q_c. Thus Q_c has a superattracting p-cycle if and only if $Q_c^p(0) = 0$.

We will find the solutions for the first values of p.

Fixed points. The equation $Q_c(0) = 0$ reduces to $c = 0$ and $Q_0(z) = z^2$ has at $z^* = 0$ a superattracting fixed point.

2-cycles. For $Q_c^2(0) = c^2 + c = 0$, the solutions are $c = 0, -1$ and then $c = -1$ is the new possible value. Also $Q_{-1}(z) = z^2 - 1$ has the superattracting 2-cycle $\{0, -1\}$.

3-cycles. For $Q_c^3(0) = c^4 + 2c^3 + c^2 + c = 0$, the equation $c^3 + 2c^2 + c + 1 = 0$ has three solutions:

$$c_1 = -1.75448, \quad c_2 = -0.12256 + 0.7448i, \quad \text{and} \quad c_3 = -0.12256 - 0.7448i.$$

Now we will study the relationship between $\mathbb{M}$ and the set $\mathbb{M}_p$ of all c such that the orbit of the origin under Q_c is periodic.

Theorem 10.3.6. *One has $\partial(\mathbb{M}) \subset \overline{\mathbb{M}_p} \subset \mathbb{M}$.*

Proof. The inclusion $\overline{\mathbb{M}_p} \subset \mathbb{M}$ is trivial. To prove that $\partial(\mathbb{M}) \subset \overline{\mathbb{M}_p}$, suppose $c_0 \in \partial(\mathbb{M})$ but $c_0 \notin \overline{\mathbb{M}_p}$. As $0 \notin \partial(\mathbb{M})$, it follows that $c_0 \neq 0$. Then there exists a disc $\mathbb{D}(c_0)$ such that

$$\mathbb{D}(c_0) \cap \overline{\mathbb{M}_p} = \emptyset \quad \text{and} \quad 0 \notin \mathbb{D}(c_0).$$

Let $g(c)$ be a uniform and analytic branch of $\sqrt{-c}$ defined on $\mathbb{D}(c_0)$. Then consider the functions

$$f_n(c) = \frac{Q_c^n(0)}{g(c)} \quad (c \in \mathbb{D}(c_0)).$$

For every $n \in \mathbb{N}$, none of the values 1 and 0 are attained in $\mathbb{D}(c_0)$ by f_n. Indeed, note that the statements:
(1) $Q_c^n(0) = \sqrt{-c}$ for some $n \in \mathbb{N}$.
(2) $Q_c^{n+1}(0) = 0$ for some $n \in \mathbb{N}$.
(3) 0 is periodic for Q_c

are equivalent. So, by Montel's theorem, (f_n) is normal in $\mathbb{D}(c_0)$. Thus, if $c_1 \notin \mathbb{M}$, there exists a divergent subsequence of $(Q_{c_1}^n(0))_n$ that we denote by $(Q_{c_1}^{n_k}(0))_k$. It is clear that $(f_{n_k}(c_1))_k$ is also divergent. By normality, $(f_{n_k})_k$ admits a subsequence uniformly divergent on compact subsets of $\mathbb{D}(c_0)$, but this is not possible for $c \in \mathbb{D}(c_0) \cap \mathbb{M}$. □

Finally, we point out the following dichotomy theorem. The proof may be consulted in [20].

Theorem 10.3.7. *If $c \in \mathbb{M}$, the Julia set of Q_c is connected. Otherwise, the Julia set of Q_c is totally disconnected.*

10.4 The parameter plane of the exponential family

We know that the origin is the only singular point of the family $E_\lambda(z) = \lambda e^z$, where $\lambda \in \mathbb{C}$. Then the parameter plane of the family is $\mathbb{P} = \{\lambda \in \mathbb{C} : (E_\lambda^n(0))_n \text{ is bounded}\}$. We will see that there are some similarities with the Mandelbrot set, but also important differences. First, we determine the set $C(1)$. If z is an attracting fixed point for E_λ with multiplier ζ, then

$$\lambda e^z = z, \quad E_\lambda'(z) = \lambda e^z = \zeta, \quad \text{and} \quad |\zeta| < 1.$$

Hence $\zeta = z$, and expressing λ in terms of the multiplier ζ yields

$$\lambda = \zeta e^{-\zeta} \quad (|\zeta| < 1). \tag{10.8}$$

Then $C(1)$ is the region limited by the cardioid C whose equation is $r = e^{-\cos\theta}$, $\theta \in [0, 2\pi]$.

While the sets $C(k)$ for the quadratic family are bounded, they are unbounded for the exponential family for $k \geq 2$. In the proof of this result, we will use the notations and ideas of Section 10.2. Recall that if $\lambda \in C(k)$, $z^*(\lambda)$ is the k-periodic attracting point of E_λ towards which the singular orbit $(E_\lambda^{kn}(0))_n$ converges.

Proposition 10.4.1. *If $k \geq 2$, every connected component of $C(k)$ is unbounded.*

Proof. First, we will prove the following:

Claim. If $\lambda \in C(k)$, then

$$|z^*(\lambda)| \leq \max\{|E_\lambda^j(z)| : z \in \mathbb{D}, j = 1, \ldots, k-1\}.$$

Indeed, given an attracting k-cycle $z_0^*, z_1^*, \ldots, z_{k-1}^*$ for E_λ, by the chain rule and the fact that $E_\lambda'(z_j^*) = z_{j+1}^*$, we have

$$|(E_\lambda^k)'(z_0^*)| = \prod_{j=0}^{j=k-1} |E_\lambda'(z_j^*)| = \prod_{j=0}^{j=k-1} |z_{j+1}^*| < 1.$$

Therefore, there is at least one z_j^* for which $|z_j^*| < 1$. As $z^*(\lambda) = E_\lambda^s(z_j^*)$ for some $s = 0, 1, \ldots, k-1$, the proof of the claim concludes.

Now, by contradiction, suppose C is a bounded connected component of $C(k)$. We will prove that $\chi(C)$ is open and closed in $\mathbb{D}$ and, therefore, $\chi(C) = \mathbb{D}$. However, note that $\chi(\lambda) = 0$ if and only if $\lambda = 0$ and $0 \in C_1$; therefore, $0 \notin \chi(C)$ and we get a contradiction. As χ is analytic in C, $\chi(C)$ is open. Then the proof concludes if we show that $\chi(C)$ is closed in $\mathbb{D}$. Let $w_0 \in \mathbb{D}$ and $\chi(\lambda_n) \to w_0$ with $\lambda_n \in C$ for all $n \in \mathbb{N}$. According to the claim, $(z^*(\lambda_n))$ is bounded, thus there exists a sequence $n_p \to \infty$ such that $(z^*(\lambda_{n_p}))_p \to z \in \mathbb{C}$. By the boundedness of C, there is a convergent subsequence of $(\lambda_{n_p})_p$. For the sake of simplicity, we go on denoting it by $(\lambda_{n_p})_p$. If $\lambda \in \mathbb{C}$ denotes its limit, then

$$\lambda = \lim_{p\to\infty} \lambda_{n_p} \quad \text{and} \quad z = \lim_{p\to\infty} z^*(\lambda_{n_p}). \tag{10.9}$$

On the other hand, for $p \in \mathbb{N}$, we have

$$|(E_{\lambda_{n_p}}^k)'(z^*(\lambda_{n_p}))| < 1 \quad \text{and} \quad E_{\lambda_{n_p}}^k(z^*(\lambda_{n_p})) = z^*(\lambda_{n_p}).$$

Then passing to the limit as $p \to \infty$, we deduce that

$$|(E_\lambda^k)'(z)| \leq 1 \quad \text{and} \quad E_\lambda^k(z) = z. \tag{10.10}$$

As $\chi(\lambda_{n_p}) = (E_{\lambda_{n_p}}^k)'(z^*(\lambda_{n_p}))$ for all $p \in \mathbb{N}$, it follows that $w_0 = (E_\lambda^k)'(z)$. Thus $|(E_\lambda^k)'(z)| < 1$. It is easy to prove that k is the period of z and, in consequence, $\lambda \in C(k)$. Finally, take a connected neighborhood V of λ contained in $C(k)$. For p sufficiently large, $\lambda_{n_p} \in V$ and the maximality of C yields $V \subset C$. Therefore, $\lambda \in C$ and $\chi(\lambda) = \lim_{p\to\infty} \chi(\lambda_{n_p}) = w_0$. This shows that $w_0 \in \chi(C)$. □

In the next chapter, we show an interesting relationship between the parameter planes of the families $E_\lambda(z) = \lambda e^z$ and $P_{N,\lambda}(z) = \lambda(1 + z/N)^N$, where $N \in \mathbb{N}$ $(N \geq 2)$ and $\lambda \in \mathbb{C}$.

10.5 Exercises

1. Let $F_\lambda(z) = \lambda z e^z$ and let $C(1)$ be the set of all complex numbers λ such that F_λ has an attracting fixed point. Determine the set $C(1)$.
2. Prove that the intersection of $\mathbb{M}$ with the real axis is the interval $[-2, 1/4]$.
3. Prove that

$$|Q_c^n(0)| \geq |c|(|c| - 1)^{2^{n-2}},$$

for all $|c| > 2$ and $n \geq 2$. Deduce that $Q_c^n(0) \to \infty$ for all $|c| > 2$.
4. Show that $c \in \mathbb{M}$ for all $|c| \leq 1/4$.
5. If c_0 is a complex constant such that Q_{c_0} has an attracting m-periodic point z_0, then c_0 belongs to the interior of $\mathbb{M}$.
6. Consider the sequence of functions (g_n) defined by $g_n(c) = Q_c^n(0)$. Show that (g_n) is normal in a neighborhood of c_0 if and only if $c_0 \in \partial(\mathbb{M})$.
7. Consider the logistic complex family $P_\mu(z) = \mu z(1 - z)$. Determine the set C_2 of all μ such that P_μ has an attracting 2-periodic orbit.
8. If $\lambda \in C_k$ for some $k \in \mathbb{N}$, we know that E_λ has a unique attracting k-periodic orbit. Show that the Fatou set of E_λ coincides with its attractive basin.
9. Prove that the set formed by all c such that Q_c has an attracting 2-cycle is the disc $\mathbb{D}_{1/4}(-1)$.
10. Let $C \subset C_k \subset \mathbb{M}$ be a hyperbolic component. Show that $(Q_c^{nk}(0))_n$ is uniformly convergent on compact subsets of C.

Bibliography remarks

In Section 10.2, inspired by [27, pp. 181–206], we consider an arbitrary family f_c satisfying conditions C1–C3. This fact seems not to have been considered in the literature.

Section 10.3 is based on [20, Section VIII.2].

In Section 10.4, we follow [27, pp. 181–206].

11 Convergence and dynamics

In this final chapter, we tackle the following question: If f_n and f are entire functions such that $f_n \to f$ uniformly on compact subsets of $\mathbb{C}$, what can we say about the dynamics of f from those of f_n? In general, the dynamics may be completely different even if f and f_n are very close. However, some information could be obtained in specific cases. For instance, Devaney, Goldberg, and Hubbard [29] studied the exponential family $E_\lambda(z)$ by approximating with the sequence of polynomials $P_{n,\lambda}(z) = \lambda(1 + z/n)^n$. Krauskopf [56] proved that $\mathbb{J}(P_{n,\lambda})$ converges to $\mathbb{J}(E_\lambda)$ in the Hausdorff metric under certain conditions. We start the chapter introducing the Hausdorff metric and reviewing the properties that we will need.

11.1 The Hausdorff metric

Let (X, d) be a metric space and denote by $\mathrm{Comp}(X)$ the set of all nonempty compact subsets of X.

Recall that, for each subset B of X, $d(a, B)$ is defined by

$$d(a, B) = \inf\{d(a, b) : b \in B\},$$

for all $a \in X$. At first sight, the distance between two sets A and B belonging to $\mathrm{Comp}(X)$ could be given by

$$d(A, B) = \sup\{d(a, B) : a \in A\}.$$

With this definition, in general, $d(A, B) \neq d(B, A)$.

Definition 11.1.1. For all A and B belonging to $\mathrm{Comp}(X)$, the Hausdorff distance between A and B is defined by

$$D(A, B) = \max\{d(A, B), d(B, A)\}.$$

There is a more convenient form of representing the Hausdorff distance:

$$D(A, B) = \inf\{\varepsilon > 0 : A \subset N_\varepsilon(B), B \subset N_\varepsilon(A)\},$$

where $N_\varepsilon(A) = \{x \in X : \text{there is } a \in A \text{ such that } d(x, a) < \varepsilon\}$. If X is complete, it may be proved that $(\mathrm{Comp}(X), D)$ is a complete metric space (see, [43]).

Theorem 11.1.2. *If $F : X \to X$ is a contraction, the map $\overline{F} : A \in \mathrm{Comp}(X) \to F(A) \in \mathrm{Comp}(X)$ is also a contraction.*

Proof. Recall that, for each $a \in X$ and $B \in \mathrm{Comp}(X)$, there exists $b \in B$ such that $d(a, b) = d(a, B)$. Given A and B in $\mathrm{Comp}(X)$, for each $a \in A$ choose $b(a) \in B$ such that $d(a, B) = d(a, b(a))$. Then

https://doi.org/10.1515/9783111689685-011

$$d(F(a), F(b(a))) \leq k\, d(a, b(a)) = k\, d(a, B) \leq k\, d(A, B) \leq k\, D(A, B),$$

which yields $d(F(a), F(B)) \leq k\, D(A, B)$ for every $a \in A$ and, in consequence, we have proved that $D(F(A), F(B)) \leq k\, D(A, B)$. □

Finally, following Douady's ideas [33], we may decompose the continuity property for Comp(X)-valued maps into upper and lower semicontinuity properties. First, we recall the case of real-valued functions.

Let Λ be a (metrizable) topological space and let $\Phi : \Lambda \to \mathbb{R}$ be a map. We say that Φ is *upper semicontinuous* at $\lambda_0 \in \Lambda$ if, given any $\varepsilon > 0$, there exists a neighborhood V of λ_0 such that

$$\Phi(\lambda) < \Phi(\lambda_0) + \varepsilon \quad \text{for all } \lambda \in V,$$

and *lower semicontinuous* if

$$\Phi(\lambda) > \Phi(\lambda_0) - \varepsilon \quad \text{for all } \lambda \in V.$$

By analogy, a map defined from Λ into Comp(X) is called upper semi-continuous at $\lambda_0 \in \Lambda$ if, given $\epsilon > 0$, there exists a neighborhood V of λ_0 such that

$$\Phi(\lambda) \subset N_\varepsilon(\Phi(\lambda_0)) \quad \text{for all } \lambda \in V,$$

and lower semicontinuous if

$$\Phi(\lambda_0) \subset N_\varepsilon(\Phi(\lambda)) \quad \text{for all } \lambda \in V.$$

Nevertheless, in the case of Comp(X)-valued maps, the two semicontinuity properties play very different roles. As the next proposition shows, the upper semicontinuity is more natural.

If X and Y are metrizable topological spaces, we recall that a continuous map $f : X \to Y$ is said to be *proper* if the preimage of a compact set in Y is compact in X. Every proper map is closed, i. e., the image of a closed set is closed.

Given a map $\Phi : \Lambda \to \text{Comp}(X)$, we denote by Π the set defined by

$$\Pi = \{(\lambda, x) \in \Lambda \times X : x \in \Phi(\lambda)\}.$$

Proposition 11.1.3. *Under the above conditions, the following statements are equivalent:*
(i) *Φ is upper semicontinuous.*
(ii) *The set Π is closed in $\Lambda \times X$ and the projection map $p_\Pi : \Pi \to \Lambda$ is proper.*

Proof. We only prove the implication (ii) $\Longrightarrow$ (i) and leave the converse (i) $\Longrightarrow$ (ii) as Exercise 1. For fixed λ_0 and $\varepsilon > 0$, consider the set $M = \{(\lambda, x) \in \Pi : D(x, \Phi(\lambda_0)) \geq \varepsilon\}$. This set is closed in Π and does not intersect $p_{\Pi}^{-1}(\lambda_0)$. As $p_{\Pi}(M)$ is closed in Λ, its complement W is an open neighborhood of λ_0. Finally, note that $d(\Phi(\lambda), \Phi(\lambda_0)) < \varepsilon$. □

Proposition 11.1.4. *If Λ is locally compact, then condition (ii) is equivalent to:*
(ii') *Π is closed in $\Lambda \times X$ and, for each $\lambda_0 \in \Lambda$, there exist a neighborhood V of λ_0 and a compact set $K \subset X$ such that $\Phi(\lambda) \subset K$ for all $\lambda \in V$.*

Proof. (ii') $\Longrightarrow$ (ii). We only need to prove that p_{Π} is proper. Given any compact subset $L \subset \Lambda$, let $((\lambda_n, x_n))$ be a sequence in $p_{\Pi}^{-1}(L)$. Then $\lambda_n = p_{\Pi}(\lambda_n, x_n) \in L$ for all $n \in \mathbb{N}$. Thus (λ_n) admits a subsequence (λ_{n_k}) which is convergent to certain $\lambda_0 \in L$. By hypothesis, there exist a neighborhood V of λ_0 and a compact set $K \subset X$ such that

$$\Phi(\lambda) \subset K \quad \text{for all } \lambda \in V.$$

Taking V sufficiently small, we may suppose that $\lambda_n \in V$ for all n. As $x_n \in \Phi(\lambda_n) \subset K$, there is a subsequence of $(x_{n_k})_k$ convergent to some $x_0 \in K$. To simplify, we may assume that the whole sequence $(x_{n_k})_k$ is convergent to x_0 and, in consequence, $(\lambda_{n_k}, x_{n_k}) \to (\lambda_0, x_0)$. Since Π is closed in $\Lambda \times X$, it follows that $(\lambda_0, x_0) \in \Pi$. Finally, note that $(\lambda_0, x_0) \in p_{\Pi}^{-1}(L)$.

The proof of (ii) $\Longrightarrow$ (ii') is left to the reader as Exercise 6. □

We finish this section with a beautiful result related to the Julia set of polynomials. Let $\mathbb{P}_d$ be the set of complex polynomials of degree $d \geq 2$,

$$P(z) = a_d z^d + a_{d-1} z^{d-1} + \cdots + a_1 z + a_0,$$

with $a_d \neq 0$. Then $\mathbb{P}_d$ may be identified with $\mathbb{C}^* \times \mathbb{C}^d$. Recall that $\mathbb{B}(f)$ denotes the filled Julia set of f.

Theorem 11.1.5. (i) *The map $P \in \mathbb{P}_d \to \mathbb{B}(P) \in \mathrm{Comp}(\mathbb{C})$ is upper semicontinuous.*
(ii) *The map $P \in \mathbb{P}_d \to J(P) \in \mathrm{Comp}(\mathbb{C})$ is lower semicontinuous.*

Proof. (i) We leave the proof as Exercise 2.
(ii) We know that, for every polynomial $P(z)$, $\mathbb{J}(P)$ is the closure of the set of repelling periodic points. Given $P_0 \in \mathbb{P}_d$ and $\varepsilon > 0$, choose a finite set $X = \{z_1, z_2, \ldots, z_p\}$ of repelling periodic points for P_0 such that $\mathbb{J}(P_0) \subset \cup_k \mathbb{D}_{\varepsilon/2}(z_k)$ and, in consequence,

$$d(\mathbb{J}(P_0), X) \leq \varepsilon/2. \tag{11.1}$$

Note that each z_k is a solution of the equation $F_k(P, z) = P^{p_k}(z) - z = 0$, with p_k being the period of z_k. Since $(P_0^{p_k})'(z_k) - 1 \neq 0$, we can apply the implicit function theorem to determine a neighborhood V of P_0 in $\mathbb{P}_d$ and analytic functions $a_k(P)$ defined in V satisfying $F_k(P, a_k(P)) = 0$ for all $P \in V$ and $k = 1, 2, \ldots, p$. In other words, we have $P^{p_k}(a_k(P)) = a_k(P)$ for all $P \in V$ and $k = 1, \ldots, p$. Then $a_k(P)$ is p_k-periodic for P and $a_k(P_0) = x_k$. If we choose V small enough, then $|a_k(P) - x_k| \leq \varepsilon/2$ and the points

$a_k(P)$ are repelling for every $P \in V$ and $k = 1, \ldots, p$. Thus $a_k(P)$ belongs to $\mathbb{J}(P)$ for all $P \in V$ and $k = 1, \ldots, p$. From this, obviously, we deduce that

$$d(X, \mathbb{J}(P)) \leq d(X, \{a_k(P) : k = 1, \ldots, p\}) \leq \frac{\varepsilon}{2}.$$

Combining this inequality with (11.1), the proof concludes. □

Corollary 11.1.6. *Let $P_0 \in \mathbb{P}_d$ be such that the interior of the filled Julia set is empty. Then $\mathbb{J}(P_0) = \mathbb{B}(P_0)$, and the map $P \in \mathbb{P}_d \to J(f) \in \mathrm{Comp}(\mathbb{C})$ is continuous at P_0.*

11.2 Convergence of attracting orbits

If the sets A_n and A are k-periodic orbits, then we have the following simple characterization of the convergence $A_n \to A$ in the Hausdorff metric.

Proposition 11.2.1. *Let $f : \mathbb{C} \to \mathbb{C}$ be continuous and let $O^+(z_0)$ and $(O^+(z_n))$ be k-periodic orbits. Then $O^+(z_n)$ converges to $O^+(z_0)$ in the Hausdorff metric if and only if there is a point $\hat{z}$ in $O^+(z_0)$ with the following property: for all $\varepsilon > 0$, there exists an $N \in \mathbb{N}$ such that $O^+(z_n) \cap \mathbb{D}(\hat{z}, \varepsilon) \neq \emptyset$ for $n \geq N$.*

Proof. The necessity is obvious. To prove the sufficiency, we only give a sketch of the proof and leave the minor details to the reader. For a fixed $\varepsilon > 0$, choose $0 < \delta_1 < \delta_2 < \cdots < \delta_{k-1} < \varepsilon$ such that

$$\begin{cases} f^{k-1}(\mathbb{D}(z, \delta_{k-1})) \subset \mathbb{D}(f^{k-1}(z), \varepsilon), \\ f^{k-2}(\mathbb{D}(z, \delta_{k-2})) \subset \mathbb{D}(f^{k-2}(z), \delta_{k-1}), \\ \quad \vdots \\ f^{2}(\mathbb{D}(z, \delta_{2})) \subset \mathbb{D}(f^{2}(z), \delta_{3}), \\ f(\mathbb{D}(z, \delta_{1})) \subset \mathbb{D}(f(z), \delta_{2}), \end{cases} \tag{11.2}$$

for all $z \in O^+(z_0)$. Notice that we only need the continuity of $f, f^2, \ldots, f^{k-1}$ at every point of the finite set $O^+(z_0)$. By hypothesis, there exists an $N \in \mathbb{N}$ such that $M_n = O^+(z_n) \cap \mathbb{D}(\hat{z}, \delta_1) \neq \emptyset$ for all $n \geq N$. Then, taking $\hat{w}_n$ in M_n, for all $n \geq N$, and applying (11.2), we deduce that $O^+(z_n) \subset \bigcup_{z \in O^+(z_0)} \mathbb{D}_\varepsilon(z) = N_\varepsilon(O^+(z_0))$ for $n \geq N$. □

Lemma 11.2.2. *Let f and f_n $(n = 1, 2, \ldots)$ be entire functions. If $f_n \to f$ uniformly on compact subsets of $\mathbb{C}$, then $f_n^k \to f^k$ uniformly on compact sets for every $k \in \mathbb{N}$.*

Proof. To simplify, we will prove that $f_n^2 \to f^2$ uniformly on $\overline{\mathbb{D}}_R(0)$ for every $R > 0$. Given R and $\varepsilon > 0$, put $M = \max\{|f(z)| : |z| \leq R\}$. Due to the uniform convergence, there is an n_1 such that

$$|f_n(z)| \leq M + 1 \quad \text{for all } |z| \leq R \text{ and } n \geq n_1. \tag{11.3}$$

As f is uniformly continuous on the compact $\overline{\mathbb{D}}_{M+1}(0)$, there exists $\delta > 0$ such that

$$\left.\begin{array}{l} u, v \in \overline{\mathbb{D}}_{M+1}(0) \\ |u - v| < \delta \end{array}\right\} \quad \Longrightarrow \quad |f(u) - f(v)| < \frac{\varepsilon}{2}. \tag{11.4}$$

Again by the uniform convergence, this time on $\overline{\mathbb{D}}_{M+1}(0)$, there is an n_2 for which

$$|f_n(z) - f(z)| < \frac{\varepsilon}{2} \quad \text{for } z \in \overline{\mathbb{D}}_{M+1}(0),\ n \geq n_2. \tag{11.5}$$

By (11.3)–(11.5), for $n \geq \max\{n_j : j = 1, 2\}$ and $\zeta \in \overline{\mathbb{D}}_R(0)$, we have

$$|f_n^2(\zeta) - f^2(\zeta)| \leq |f_n(f_n(\zeta)) - f(f_n(\zeta))| + |f(f_n(\zeta)) - f(f(\zeta))| \leq \varepsilon. \qquad \square$$

Theorem 11.2.3. *Let f and f_n be entire functions such that $f(z) = \lim_{n\to\infty} f_n(z)$ uniformly on compact subsets of $\mathbb{C}$. If z_0^* is an attracting (repelling) k-periodic point for f, then there exists an $N \in \mathbb{N}$ such that f_n has an attracting (repelling) k-periodic point z_n for $n \geq N$ and the orbits $O^+(z_n)$ converge to $O^+(z_0^*)$ in the Hausdorff metric.*

Proof. Suppose z_0^* is attracting and choose $\varepsilon > 0$ such that $|(f^k)'(z_0^*)| < 1 - \varepsilon$. There exists a disc $\mathbb{D}_r(z_0^*)$ satisfying $|(f^k)'(z)| < 1 - \varepsilon$ for $z \in \mathbb{D}_r(z_0^*)$. As $(f_n^k)'$ converges uniformly to $(f^k)'$ on $\overline{\mathbb{D}}_r(z_0^*)$, there is an $N \in \mathbb{N}$ such that $|(f_n^k)'(z) - (f^k)'(z)| < \varepsilon$ for $n \geq N$ and $z \in \overline{\mathbb{D}}_r(z_0^*)$. Then, for $n \geq N$ and $z \in \mathbb{D}_r(z_0^*)$, we have

$$|(f_n^k)'(z)| < |(f_n^k)'(z) - (f^k)'(z)| + |(f^k)'(z)| < 1. \tag{11.6}$$

Since $f^k(z) - z$ is not identically null, we may choose r sufficiently small such that z_0^* is the unique zero of $f^k(z) - z$ in $\overline{\mathbb{D}}_r(z_0^*)$. Hurwitz's theorem asserts that there exists an $N' \in \mathbb{N}$ $(N' > N)$ such that $f_n^k(z) - z$ has a unique zero z_n in $\mathbb{D}_r(z_0^*)$ for every $n \geq N'$. Thus z_n is a periodic point for f_n, which is attracting in view of (11.6). Now we have to prove that $O^+(z_n) \to O^+(z_0^*)$. According to Proposition 11.2.1, we must show that there is some $\hat{z} \in O^+(z_0^*)$ such that, for every $\varepsilon > 0$, there exists an $N \in \mathbb{N}$ satisfying

$$\mathbb{D}_\varepsilon(\hat{z}) \cap O^+(z_n) \neq \emptyset \quad \text{for } n \geq N.$$

We consider the point $\hat{z} = f(z_0^*) \in O^+(z_0^*)$. Given $\varepsilon > 0$, there is $0 < s < r$ satisfying $f(\mathbb{D}_s(z_0^*)) \subset \mathbb{D}(\hat{z}, \varepsilon/2)$. On the other hand, as $f_n \to f$ uniformly on $\mathbb{D}_s(z_0^*)$, there is an $N_1'' > N''$ such that $|f_n(z) - f(z)| < \varepsilon/2$ for $n \geq N_1''$ and $z \in \mathbb{D}_s(z_0^*)$. Then, for $n \geq N_1''$, we have

$$|f_n(z_n) - f(z_0^*)| \leq |f_n(z_n) - f(z_n)| + |f(z_n) - f(z_0^*)| < \varepsilon,$$

and, therefore, $f_n(z_n) \in \mathbb{D}_\varepsilon(\hat{z}) \cap O^+(z_n)$ for $n \geq N_1''$.

Finally, we must prove that k is the period of z_n. Let $O^+(z_0^*) = \{z_0^*, z_1^*, \ldots, z_{k-1}^*\}$ and $\varepsilon < (1/2)\min\{d(z_i^*, z_j^*) : 0 \leq i, j \leq k - 1 (i \neq j)\}$. Since $O^+(z_n) \to O^+(z_0^*)$, there exists an

$N \in \mathbb{N}$ such that $\mathbb{D}_\varepsilon(z_i^*) \cap O^+(z_n) \neq \emptyset$ for $i = 0, 1, \ldots, k-1$ and $n \geq N$. This proves that the orbits $O^+(z_n)$, for $n \geq N$, contain k different points. □

Suppose f_n ($n = 1, 2, \ldots$) and f only have one singular value and denote by $C_{n,k}$ the set of all $\lambda \in \mathbb{C}$ such that f_n has an attracting orbit of period k. The next corollary follows from Theorem 11.2.3.

Corollary 11.2.4. *If $\lambda \in C_k$, then there exists an $N_0 \in \mathbb{N}$ such that $\lambda \in C_{n,k}$ for $n \geq N_0$.*

Theorem 11.2.5. *Let f and f_n be entire functions such that $f(z) = \lim_{n\to\infty} f_n(z)$ uniformly on compact subsets of $\mathbb{C}$, and let z_0 be an attracting k-periodic point for f. Then, for every $z \in \Omega(f^k, z_0)$ and any neighborhood $U(z)$ of z with $\overline{U}(z) \subset \Omega(f^k, z_0)$, there is an $N \in \mathbb{N}$ such that $U(z) \subset \mathbb{F}(f_n)$ for all $n \geq N$.*

Proof. For the sake of simplicity, we suppose z_0 is a fixed point of f. The proof may be easily extended to the case of a periodic point by considering the kth iterate of f.

First, we prove that there exist a neighborhood $V(z_0)$ of z_0 and $N \in \mathbb{N}$ such that $V(z_0) \subset \Omega(f, z_0) \cap \mathbb{F}(f_n)$ for all $n \geq N$. Indeed, since z_0 is attracting, there is a disc $\mathbb{D}_r(z_0) \subset \Omega(f, z_0)$ such that

$$|(f)'(z)| \leq \alpha < 1 \quad \text{for all } z \in \mathbb{D}_r(z_0).$$

By uniform convergence, there exist $N \in \mathbb{N}$, $\alpha_0 \in (0,1)$, and $r^* < r$ satisfying

$$|f_n'(z)| \leq \alpha_0 < 1 \quad \text{for all } z \in \mathbb{D}_{r^*}(z_0),\ n \geq N. \tag{11.7}$$

Theorem 11.2.3 asserts that there is an $N' \geq N$ such that, for $n \geq N'$, each f_n has an attracting fixed point z_n satisfying $z_n \to z_0$. So, taking N' large enough, we may assume that $z_n \in \mathbb{D}_{r^*}(z_0)$ for all $n \geq N'$. By (11.7), we may choose $0 < s < r^*$ and $N'' > N'$ such that $f_n(\mathbb{D}_s(z_0)) \subset \mathbb{D}_s(z_0)$ for every $n \geq N''$. Again by (11.7), we have

$$|f_n^k(z) - z_n| = \left| \int_{f_n^{k-1}(z_n)}^{f_n^{k-1}(z)} f_n'(\zeta)\, d\zeta \right| \leq \alpha_0 |f_n^{k-1}(z_n) - f_n^{k-1}(z)|,$$

for all $z \in \mathbb{D}_s(z_0)$, $n \geq N''$ and $k \in \mathbb{N}$. Now, by induction, we get

$$|f_n^k(z) - z_n| \leq \alpha_0^k |z - z_n|.$$

This proves that $\mathbb{D}_s(z_0) \subset \Omega(f_n, z_n)$ for all $n \geq N''$.

Finally, let $U(z)$ be a neighborhood of z with $\overline{U}(z) \subset \Omega(f, z_0)$ (we may suppose that $U(z)$ is bounded). As $f^n(\zeta) \to z_0$ uniformly on the compact $\overline{U}(z)$, there exists $p \in \mathbb{N}$ such that $f^p(U(z)) \subset \mathbb{D}_{s/2}(z_0)$. On the other hand, $f_n^p \to f^p$ locally uniformly in $\mathbb{C}$ and, therefore, there is an $N > N''$ such that $f_n^p(U(z)) \subset \mathbb{D}_s(z_0)$ for $n \geq N$. The result follows from the invariance of attraction basins. □

11.3 Convergence of Julia sets

In this section, we view Julia sets as compact subsets of $\hat{\mathbb{C}}$ in a natural way. In fact, we may compactify $\mathbb{J}(f)$ by assuming that the point ∞ belongs to $\mathbb{J}(f)$ or, in other words, considering the set $\hat{\mathbb{J}}(f) = \mathbb{J}(f) \cup \{\infty\}$.

In general, the relationship between the $\mathbb{J}(f_n)$'s, for n sufficiently large, and $\mathbb{J}(f)$ may be very weak as the following example shows.

Example 11.3.1. Let $f(z) = (1/e)e^z$ and $f_n(z) = \lambda_n e^z$, where $(\lambda_n) \to 1/e$ is a sequence of real numbers such that $\lambda_n > 1/e$ for all $n \in \mathbb{N}$. By Corollary 9.6.2, $\mathbb{J}(f_n) = \mathbb{C}$ for all n, nevertheless, $\mathbb{F}(f)$ is the basin of attraction of the neutral fixed point $z = 1$.

Kisaka [52] proved the following theorem.

Theorem 11.3.2. *If f_n $(n = 1, 2, \dots)$ and f are entire functions such that $f_n \to f$ uniformly on compact subsets of $\mathbb{C}$, then the following statements hold:*
(i) *If $\mathbb{F}(f)$ contains only attractive basins, then $\hat{\mathbb{J}}(f_n) \to \hat{\mathbb{J}}(f)$ in the Hausdorff metric.*
(ii) *If $\mathbb{J}(f) = \mathbb{C}$, then $\hat{\mathbb{J}}(f_n) \to \hat{\mathbb{J}}(f)$ in the Hausdorff metric.*

Proof. We only prove (i).

(i) By hypothesis, for every $w \in \mathbb{F}(f)$, there exists an attracting periodic point for f satisfying

$$w \in \Omega(f, z_0) \subset \mathbb{F}(f).$$

We may suppose z_0 is a fixed point. By Hurwitz's theorem, there exists an $n_0 \in \mathbb{N}$ such that f_n has an attracting fixed point $z_{0,n}$ satisfying $z_{0,n} \to z_0$ (the argument is similar to that used in the proof of Theorem 11.2.3). Fix $\varepsilon > 0$ sufficiently small such that $\overline{\mathbb{D}}_\varepsilon(w) \subset \Omega(f, z_0)$. By Theorem 11.2.5, there is an $N \in \mathbb{N}$ such that

$$\mathbb{D}_\varepsilon(w) \subset \mathbb{F}(f_n) \quad \text{for } n \geq N. \tag{11.8}$$

Therefore, for every $w \in \hat{\mathbb{C}}/N_\varepsilon(\hat{\mathbb{J}}(f)) \subset \mathbb{F}(f)$, there exists an $N(w) \in \mathbb{N}$ such that $\mathbb{D}_\varepsilon(w) \subset \mathbb{F}(f_n)$ for all $n \geq N(w)$. Since $\hat{\mathbb{C}}/N_\varepsilon(\hat{\mathbb{J}}(f))$ is compact, there are $w_1, \dots, w_m \in \hat{\mathbb{C}}/N_\varepsilon(\hat{\mathbb{J}}(f))$ such that

$$\hat{\mathbb{C}}/N_\varepsilon(\hat{\mathbb{J}}(f)) \subset \bigcup_{j=1}^{m} \mathbb{D}_\varepsilon(w_j) \subset \mathbb{F}(f_n) \quad \text{for } n \geq N,$$

where $N = \max_j N(w_j)$. Then

$$\hat{\mathbb{J}}(f_n) \subset N_\varepsilon(\hat{\mathbb{J}}(f)) \quad \text{for } n \geq N. \tag{11.9}$$

On the other hand, by the compactness of $\hat{\mathbb{J}}(f)$, there exist $\beta_1, \dots, \beta_p \in \hat{\mathbb{J}}(f)$ such that $\hat{\mathbb{J}}(f) \subset \bigcup_{j \leq p} \mathbb{D}_c(\beta_j, \varepsilon/2)$. For each $j \leq p$, let $z_j \in \mathbb{D}(\beta_j, \varepsilon/2)$ be a repelling periodic point for f. Again by Hurwitz's theorem, there exists an $N(\beta_j) \in \mathbb{N}$ such that f_n has a repelling periodic point $z_{j,n}$ in $\mathbb{D}(\beta_j, \varepsilon/2)$ for $n \geq N(\beta_j)$. Obviously, $z_{j,n} \in \mathbb{J}(f_n)$. For each $n \geq N = \max N(\beta_j)$, we have

$$\hat{\mathbb{J}}(f) = \hat{\mathbb{J}}(f) \cap \left[\bigcup_{j=1}^{p} \mathbb{D}_c(\beta_j, \varepsilon/2) \right] = \bigcup_{j=1}^{p} [\hat{\mathbb{J}}(f) \cap \mathbb{D}_c(\beta_j, \varepsilon/2)]. \tag{11.10}$$

Finally, for every $w \in \hat{\mathbb{J}}(f) \cap \mathbb{D}_c(\beta_j, \varepsilon/2)$, we have

$$|w - z_{j,n}| \leq |w - \beta_j| + |\beta_j - z_{j,n}| < \varepsilon,$$

which yields

$$\hat{\mathbb{J}}(f) \cap \mathbb{D}_c(\beta_j, \varepsilon/2) \subset N_\varepsilon(\hat{\mathbb{J}}(f_n)).$$

Combining this with (11.10), the proof concludes. □

We finish the section considering an interesting and surprising example related to the Taylor polynomials. If f is an entire function, it is well known that

$$f(z) = \sum_{n=0}^{\infty} a_n z^n \quad \text{uniformly on compact subsets of } \mathbb{C},$$

where $\sum a_n z^n$ is the Taylor series of f at the origin.

Example 11.3.3. We consider the function $f(z) = c^3 \cos z$ and denote by $T(n)$ its Taylor polynomial of degree $2n$ given by

$$T(2n)(z) = c^3 \sum_{k=0}^{n} \frac{(-1)^k z^{2k}}{(2k)!}.$$

In Figures 11.1 and 11.2, we show the filled Julia sets of f and $T(4)$, respectively, in the window $[-2, 2] \times [-2, 2]$. They are similar even when n is so small.

Figure 11.1: Filled Julia set (blue) of $f(z) = 1.728 \cos z$.

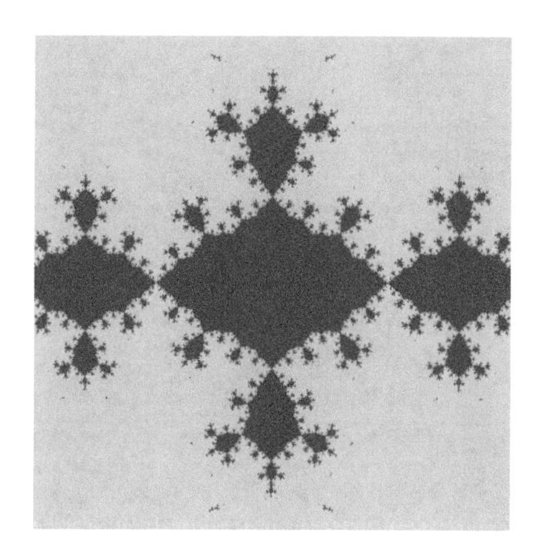

Figure 11.2: Filled Julia set (blue) of $T(4)$.

11.4 Approximation of λe^z by $\lambda(1+z/N)^N$

Now we consider the polynomials $P_{N,\lambda}(z) = \lambda(1+z/N)^N$. We know that, for each $\lambda \in \mathbb{C}$, $P_{N,\lambda}(z) \to E_\lambda(z)$ uniformly on compact subsets of $\mathbb{C}$ (Exercise 2.4). For each N, $P_{N,\lambda}(z)$ has the only critical point $z = -N$. Then $\operatorname{sing}(P_{N,\lambda}^{-1}) = \{0\}$, as in the case of the family E_λ.

Denote by $C_{N,k}$ the set of all $\lambda \in \mathbb{C}$ such that $P_{N,\lambda}(z)$ has a (finite) attracting k-periodic point. Recall that Corollary 11.2.4 asserts that there is a close relationship between C_k and $C_{N,k}$.

If E_λ has an attracting fixed point z_0, we know that $\mathbb{F}(E_\lambda) = \Omega(E_\lambda, z_0)$. Then, by Theorem 11.3.2, the following result is an obvious consequence.

Theorem 11.4.1. (i) *Let $\lambda \in \mathbb{C}$ be such that E_λ has an attracting orbit. Then $\mathbb{J}(P_{n,\lambda}) \to \mathbb{J}(E_\lambda)$ in the Hausdorff metric.*
(ii) *Let $\lambda \in \mathbb{C}$ be such that $\mathbb{J}(E_\lambda) = \mathbb{C}$. Then $\mathbb{J}(P_{n,\lambda}) \to \hat{\mathbb{C}}$ in the Hausdorff metric.*

The proof of the next theorem is Exercise 4.

Theorem 11.4.2. *Let P be the set of all $N \in \mathbb{N}$ such that $\lambda \in C_{N,k}$. If P is an infinite set and $\lambda \notin C_k$, then E_λ has a neutral k-periodic point.*

According to Theorem 6.5.1, for every polynomial $P(z)$ there exists an $R(P) > 0$ such that

$$|P^n(z)| > |P^{n-1}(z)| \quad \text{for } |z| > R(P),\ n \in \mathbb{N}$$

and $P^n(z) \to \infty$ for all z such that $|z| > R(P)$. Next, we will determine $R(N,\lambda) = R(P_{N,\lambda})$. For $|z| > 2N$, we have

$$\left|\frac{P_{N,\lambda}(z)}{z^N}\right| = |\lambda|\left|\frac{1}{N} + \frac{1}{z}\right|^N > |\lambda|\left(\frac{1}{N} - \frac{1}{2N}\right)^N = \frac{|\lambda|}{(2N)^N}.$$

Thus

$$\left|P_{N,\lambda}(z)\right| > |z|\frac{|\lambda||z|^{N-1}}{(2N)^N}.$$

Therefore, if we put $R(N,\lambda) = \max\{2N, ((2N)^N/|\lambda|)^{1/(N-1)}\}$, then

$$\left|P_{N,\lambda}(z)\right| > |z| \quad \text{for } |z| > R(N,\lambda).$$

Expressing $D = ((2N)^N/|\lambda|)^{1/(N-1)}$ in the form

$$\frac{(2N)^{N/(N-1)}}{|\lambda|^{1/(N-1)}},$$

we see that, for each λ, the numerator is greater than $2N$, while the denominator tends to 1. Then, having fixed λ, we have

$$((2N)^N/|\lambda|)^{1/(N-1)} > 2N$$

for $N > |\lambda|/2$. This yields $R(N,\lambda) = ((2N)^N/|\lambda|)^{1/(N-1)}$ for $N > |\lambda|/2$.

11.5 Exercises

1. Let $\phi : \Lambda \to \mathrm{Comp}(X)$ be an upper semicontinuous map. Show that:
   (i) If $\lambda_n \to \lambda_0$ in Λ, then $d(\phi(\lambda_n), \phi(\lambda_0)) \to 0$.
   (ii) If $\Pi = \{(\lambda, x) \in \Lambda \times X : x \in \phi(\lambda)\}$, then Π is closed in $\Lambda \times X$.
2. Suppose that $A = \lim_{n\to\infty} A_n$ in $\mathrm{Comp}(X)$. Show that A is the set of all limits of convergent subsequences $(a_{n_k})_k$ with $a_{n_k} \in A_{n_k}$.
3. Prove that the map $P \in \mathbb{P}_d \to \mathbb{B}(f) \in \mathrm{Comp}(\mathbb{C})$ is upper semi-continuous.
4. Let f and f_n be entire functions such that $f_n \to f$ uniformly on compact subsets of $\mathbb{C}$. Assume that each f_n has an attracting p-periodic point z_n^* and $z_n^* \to z^*$. Prove that:
   (i) z^* is a p-periodic point of f.
   (ii) If there exists $0 < k < 1$ such that $|(f_n^p)'(z_n^*)| \leq k$ for all $n \in \mathbb{N}$, then z^* is attracting.
   (iii) Give an example to show that the condition $|(f_n^p)'(z_n^*)| \leq 1$ for all $n \in \mathbb{N}$ does not imply that z^* is attracting.
5. Under the conditions of the last section, denote by P the set of all $N \in \mathbb{N}$ such that $\lambda \in C_{N,k}$. Show that if P is an infinite set and $\lambda \notin C_k$, then E_λ has a neutral k-periodic point.
6. Prove (ii) $\Longrightarrow$ (ii') in Theorem 11.1.4.
7. Let $\Omega \subset \mathbb{C}$ be a region and $f_n \to f$ and $g_n \to g$ locally uniformly on Ω (with all maps analytic in Ω). If $g(\Omega) \subset \Omega$, prove that $f_n \circ g_n \to f \circ g$ locally uniformly on Ω.

Bibliography remarks

Section 11.1 is based on [33, pp. 91–138].

In Section 11.3, we follow [52].

Section 11.4 is inspired by [56].

A Univalent functions in the disc

A.1 Area formula. Consequences

A function f is called univalent in a region $\Omega \subset \mathbb{C}$ if it is analytic and injective (recall that if f is univalent, then $f'(z) \neq 0$ in Ω). We will denote by $\mathbb{U}_{a,r}$ the class of all univalent functions in the disc $\mathbb{D}_r(a)$. The objective of this appendix is to prove the Koebe 1/4-theorem.

If M is the closed bounded region limited by a Jordan curve γ, the Green theorem of the classical analysis allows us to obtain the area of M through an integral along γ:

$$\text{Area}(M) = \iint_M dx\,dy = \oint_\gamma -y\,dx = \oint_\gamma x\,dy.$$

Using the complex integral along γ, the area of M may be expressed in the form

$$\text{Area}(M) = \iint_M dx\,dy = \frac{1}{2i}\oint_\gamma \bar{z}\,dz.$$

Let $K \subset \mathbb{C}$ be a connected compact set, and $\phi : \mathbb{C}/K \to \mathbb{C}/\overline{\mathbb{D}}$ univalent and surjective. Assume that

$$\phi(w) = b_1 w + b_0 + \frac{b_{-1}}{w} + \frac{b_{-2}}{w^2} + \cdots$$

is the Laurent series of ϕ, convergent for $|w| > 1$.

Theorem A.1.1 (The area formula of Gronwall). *Under the above conditions, the 2-dimensional Lebesgue measure of K is given by*

$$Area(K) = \pi \sum_{n\leq 1} n|b_n|^2.$$

Proof. For $r > 1$, let γ_r be the circle $|w| = r$ and denote by $A(r)$ the compact set limited by $\phi \circ \gamma_r$. By Green's theorem,

$$\text{Area}(A(r)) = \frac{1}{2i}\oint_{\phi\circ\gamma_r} \bar{z}\,dz.$$

If $w(\theta) = re^{i\theta}$ and $z = \sum_{n\leq 1} b_n w^n$, then

$$\text{Area}(A(r)) = \frac{1}{2i}\oint_{\phi\circ\gamma_r} \bar{z}\,dz = \frac{1}{2}\sum_{n,m} \bar{b}_n b_m m r^{n+m}\int_0^{2\pi} e^{(n-m)\theta i}\,d\theta.$$

https://doi.org/10.1515/9783111689685-012

All integrals are null except when $n = m$ and, in this case, its value is 2π. Therefore, the above equality adopts the final form

$$\operatorname{Area}(A(r)) = \pi \sum_{n\leq 1} n|b_n|^2 r^{2n}.$$

Now passing to the limit as $r \to 1$, we obtain

$$\operatorname{Area}(K) = \pi \sum_{n\leq 1} n|b_n|^2.$$ □

A trivial consequence of the above theorem is the following. As the area of K is a nonnegative number, it follows that $|b_1|^2 \geq -\sum_{n\leq 0} n|b_n|^2$ or, equivalently,

$$|b_1|^2 \geq \sum_{n=1}^{\infty} n|b_{-n}|^2. \tag{A.1}$$

Corollary A.1.2. *If $g(z) = 1/z + a_0 + a_1 z + a_2 z^2 + \cdots$ is univalent in the unit punctured disc, then*

$$\sum_{n\geq 1} n|a_n|^2 \leq 1.$$

Proof. The function $G(w) = g(1/w)$ for $w \in \mathbb{C}/\overline{\mathbb{D}}$ satisfies the conditions of Theorem A.1.1 and its Laurent series has the form

$$G(w) = w + a_0 + a_1/w + a_2/w^2 + \cdots.$$

Then, as we have just seen, $1 \geq \sum_n n|a_n|^2$ holds. □

Theorem A.1.3. *If $f(z) = z + \sum_{n=2} a_n z^n \in \mathbb{U}_{0,1}$, then $|a_2| \leq 2$.*

Proof. Expressing f in the form

$$f(z) = z(1 + a_2 z + \cdots) = z f_0(z),$$

we note that f_0 does not vanish in $\mathbb{D}$. As $\mathbb{D}$ is simply connected, there exists an analytic branch of $\sqrt{f_0(z^2)}$ defined in $\mathbb{D}$. Consider the function

$$g(z) = \frac{1}{z\sqrt{f_0(z^2)}} \quad (z \in \mathbb{D}).$$

To show that g is injective, let $z_1, z_2 \in \mathbb{D}$ be such that

$$\frac{1}{z_1\sqrt{f_0(z_1^2)}} = \frac{1}{z_2\sqrt{f_0(z_2^2)}}. \tag{A.2}$$

Obviously, we have

$$z_1^2 f_0(z_1^2) = z_2^2 f_0(z_2^2)$$

or, in terms of f,

$$f(z_1^2) = f(z_2^2).$$

Since f is injective in $\mathbb{D}$, it follows that $z_1^2 = z_2^2$, that is, $z_1 = \pm z_2$. Only the equality $z_1 = z_2$ is compatible with (A.2), so g is injective. Therefore, g satisfies the conditions of Corollary A.1.2. Now elementary calculus shows that the series of Laurent for g is

$$g(z) = 1/z - (a_2/2)z + \cdots$$

and the corollary tells us that $|a_2/2| \leq 1$. □

Theorem A.1.4 (Koebe). *If $f \in \mathbb{U}_{a,r}$, then the image $f(\mathbb{D}_r(a))$ covers the open disc of radius $|f'(a)|r/4$ with center $f(a)$, that is, we have*

$$\mathbb{D}(f(a), r|f'(a)|/4) \subset f(\mathbb{D}_r(a)).$$

Proof. To simplify the proof, we suppose that $a = f(a) = 0$ and $f'(a) = 1$. Then, the Taylor series of f is

$$f(z) = z + \sum_{n\geq 2} a_n z^n.$$

Let $c \in \mathbb{C}$ be such that f does not attain the value c in $\mathbb{D}$, and consider the function $g(z) = cf(z)/(c - f(z))$. It is easy to prove that $g \in \mathbb{U}_{0,1}$. Routine calculus shows that

$$g(z) = z + \left(a_2 + \frac{1}{c}\right)z^2 + \cdots.$$

Applying Theorem A.1.3 to f and g, we obtain

$$\frac{1}{|c|} = \left|a_2 + \frac{1}{c} - a_2\right| \leq |a_2| + \left|a_2 + \frac{1}{c}\right| \leq 2 + 2 = 4.$$ □

A.2 Exercises

1. Considering the Koebe function $f(z) = z/(1-z)^2$ for $z \in \mathbb{D}$, show that $1/4$ is optimal in Koebe's theorem.
2. If f is univalent in a region $\Omega \subset \mathbb{C}$ and $a \in \Omega$, then

$$\frac{1}{4}|f'(a)|d(a, \partial(\Omega)) \leq d(f(a), \partial(f(\Omega))) \leq 4|f'(a)|d(a, \partial(\Omega)).$$

3. Let $\Omega = (-1,1) \times (-1,1)$ and $f(z) = e^z$ for $z \in \Omega$. Show that $d(1, \partial(f(\Omega))) = 1 - e^{-1}$.
4. Let $f \in \mathbb{U}_{0,1}$ with $f(0) = 0$ and $f'(0) = 1$. For a fixed $z \in \mathbb{D}$, the Koebe transform of f is the function h defined on $\mathbb{D}$ by

$$h(w) = \frac{1}{f'(z)(1-|z|^2)}\left(f\left(\frac{z+w}{1+\bar{z}w}\right) - f(z)\right).$$

   (i) Prove that h is univalent and satisfies $h(0) = 0$ and $h'(0) = 1$.
   (ii) Applying Theorem A.1.3 to h, show that

$$\left|z\frac{f''(z)}{f'(z)} - \frac{2|z|^2}{1-|z|^2}\right| \le \frac{4|z|}{1-|z|^2}.$$

B The Newton method

B.1 The Newton method

Consider the equation $f(z) = 0$, with $f : \Omega \to \mathbb{C}$ being analytic in some region $\Omega \subset \mathbb{C}$ where $f'(z) \neq 0$. One way to find an approximate value of a zero z^* of f in Ω is the so-called Newton's method:

$$z_{n+1} = z_n - \frac{f(z_n)}{f'(z_n)}, \quad n = 0, 1, 2, \ldots,$$

where the initial value z_0 is chosen close to z^*. The sequence $\{z_0, z_1, z_2, \ldots\}$ is the orbit of z_0 under the *Newton map* $N_f(z) = z - f(z)/f'(z)$. It is obvious that looking for the zeros of $f(z)$ is equivalent to looking for the fixed points of $N_f(z)$:

$$f(z) = 0 \quad \Longleftrightarrow \quad z = z - \frac{f(z)}{f'(z)}.$$

Let us see that every zero of $f(z)$ is an attracting fixed point of $N_f(z)$. Suppose that $f(z) = (z - z^*)^m g(z)$, with $m \geq 1$ and $g(z^*) \neq 0$, then we have

$$\frac{f(z)}{f'(z)} = \frac{(z - z^*)g(z)}{(z - z^*)g'(z) + mg(z)}.$$

Notice that the right-hand side of the equality is well defined in a neighborhood of $z = z^*$, therefore, taking derivatives in the equality

$$N_f(z) = z - \frac{f(z)}{f'(z)},$$

we obtain $N_f'(z^*) = 1 - \frac{1}{m} = \frac{m-1}{m}$. As $N_f'(z^*) \in [0, 1)$, it follows that z^* is an attracting fixed point of N_f.

Thus, if z_0 is close enough to the zero z^*, Newton's method converges to z^*. Furthermore, the speed of the convergence of $z_n \to z^*$ is the best when $m = 1$, that is, when z^* is a simple zero of $f(z)$. In this case z^* is a superatractor for N_f.

B.2 The problem of Cayley

In 1879, A. Cayley proposed to extend Newton's method

$$z_{n+1} = z_n - \frac{P(z)}{P'(z)}$$

to calculate approximate values of the zeros of a complex polynomial P. He suggested that the problem should be studied globally. Concretely, he said:

https://doi.org/10.1515/9783111689685-013

"The problem consists in determining regions in the plane so that, if you take arbitrarily the initial point z_0 in one of them, then the method converges to one of the zeros of the polynomial."

In two notes published in 1879 and 1890, he faced the problem by considering $P(z) = z^2 - 1$. In this case, the zeros are $z = \pm 1$ and he proved that the attractive basins of each zero are:

$$\Omega(+1) = \{z : \operatorname{Re} z > 0\} \quad \text{and} \quad \Omega(-1) = \{z : \operatorname{Re} z < 0\}.$$

One may think that this peculiar behavior is a consequence of the simplicity of this particular polynomial $P(z)$. Nevertheless, the same is true for every polynomial of degree 2 with two different zeros. To prove it, consider the polynomial $P(z) = (z - z_1^*)(z - z_2^*)$, having z_1^* and z_2^* different. We want to obtain the attractive basins of each zero (for Newton's method). The Newton function in this case is

$$N_P(z) = z - \frac{P(z)}{P'(z)} = \frac{z^2 - z_1^* z_2^*}{2z - (z_1^* + z_2^*)}.$$

We consider the Möbius transformation $T(z) = (z - z_1^*)/(z - z_2^*)$. Elementary calculus allows us to obtain the equality

$$T \circ N(z) = T(z)^2.$$

Applying the above equality to $z = T^{-1}(w)$, we get

$$M(w) = T \circ N(z) \circ T^{-1}(w) = T(T^{-1}(w))^2 = w^2.$$

The transformation $w = T(z)$ is the key to prove the following claim:

"If the initial point z_0 is closer to z_1^*, Newton's method converges to z_1^*."

If z_0 is closer to z_1^*, then the modulus of $w_0 = T(z_0) = (z_0 - z_1^*)/(z_0 - z_2^*)$ is less than 1. As we have seen in Chapter 1, $(M^n(w_0))_n$ converges to 0. So

$$N_P^n(z_0) = T^{-1} \circ M^n \circ T(z_0) = T^{-1} \circ M^n(w_0) \to T^{-1}(0) = z_1^*.$$

In a similar way, we can prove that Newton's method converges to z_2^*, if the initial point z_0 is closer to z_2^*.

So, Cayley's idea was correct for polynomials of degree 2 with two different zeros z_1^* and z_2^*: The plane is divided in two regions by the line which passes trough $(z_1^* + z_2^*)/2$ and is orthogonal to the segment joining z_1^* to z_2^*. If we take the initial point z_0 in one of these regions, the method converges to the zero contained in it.

The above result is considered as the first result in the global theory of complex iteration.

B.3 The cubic equation

On the other hand, if we consider polynomial functions of degree 3, it is easy to show that there may be regions with positive measure where Newton's method does not converge. For this, it suffices to take a polynomial whose Newton's function has, for example, an attracting 2-cycle $\{z_0^*, z_1^*\}$. In this case, there exist two regions Ω_k $(k = 0, 1)$ such that $z_k^* \in \Omega_k$ and, choosing the initial point z_0 in one of them, the method will oscillate. In fact, the even iterates converge to one of the points of the cycle and the odd iterates converge to the other. In the next example, we determine a polynomial of degree 3 satisfying that condition.

Example B.3.1. Elementary calculus shows that the polynomial $f(z) = z^3 + az^2 + bz + c$ has the property that $\{0, 1\}$ is a 2-cycle of its Newton's function $N(z)$ whenever $a = c - 2$, $b = -c$ for all $c \neq 1$. It is easy to see that this cycle is superattracting if we take $c = -1$ or $c = 2$ since for these values $N'(0) \cdot N'(1) = 0$. This yields two possible polynomials: $f_1(z) = z^3 - 3z^2 + z - 1$ and $f_2(z) = z^3 - 2z + 2$.

B.4 Examples

As the boundaries of the convergence regions to each root are contained in the Julia set of Newton's function, the "fractal" nature of this set is the reason of the intricate geometry of those regions. Therefore, if we apply Newton's method to calculate an approximate root of the equation $f(z) = 0$, the convergence basin often presents an intricate geometry and computers allow us to obtain beautiful pictures. We will see two examples of this.

(1) The equation $ze^z - \cos(z) = 0$ has infinitely many real roots. The closest to the origin are: $z^* = -1.8640\ldots$ and $w^* = 0.5178\ldots$.
(2) The equation $e^{z^2} = 2\cos(z)$ has two real solutions $z = \pm 0.67031\ldots$.

In Figures B.1 and B.2, the regions of convergence towards each of them are colored red and yellow. If, for the initial point z_0, Newton's method does not converge to any of the roots we are considering, then we draw a black point at z_0.

At the end of Appendix C, we show another couple of examples of this type (Figures C.19 and C.20).

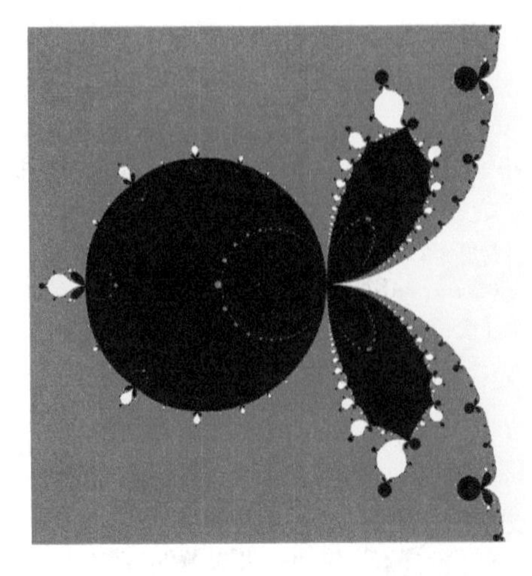

Figure B.1: In red and yellow, convergence regions towards each of two roots of $ze^z - \cos(z) = 0$.

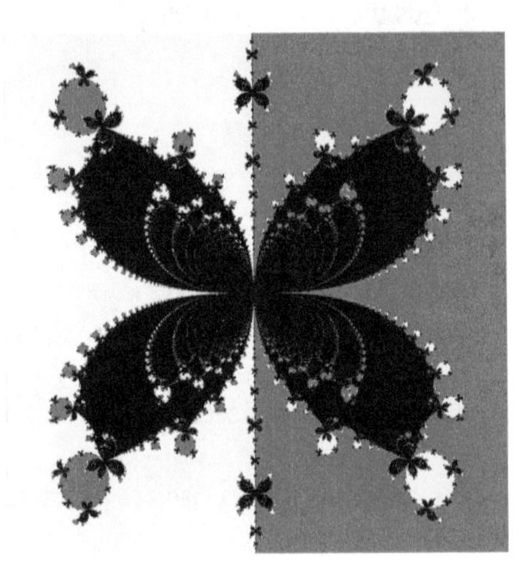

Figure B.2: In red and yellow, convergence regions towards each of two roots of $e^{z^2} - 2\cos(z) = 0$.

B.5 Exercises

1. Consider the polynomial $P(z) = (1/2)z^3 - z + 1$. Show that the inflection point $z = 0$ is a superattracting 2-periodic point, but not a fixed point.
2. Describe the dynamics of Newton's function for any polynomial of degree m with one zero of multiplicity m.
3. Let $f(z) = z^3 - 1$ and let (z_n) be the sequence of iterations obtained by applying Newton's method. Find an explicit value of r such that $z_n \to 1$ for all $|z - 1| < r$.
4. Let f be analytic in a region Ω and let $z^* \in \Omega$ be a root of f. Show that the convergence of Newton's method is of second order if the root is simple, and of first order if z^* is a multiple root. An iterative method is said to be of order $k > 0$ if there is a positive constant α so that
$$\lim_{n\to\infty} \frac{|z_{n+1} - z^*|}{|z_n - z^*|^k} = \alpha.$$

5. Show that the affine change $w = T(z) = az + b$ transforms the zeros of P without producing qualitative changes in the dynamics of Newton's function. Concretely, prove that $T^{-1} \circ N_P \circ T = N_Q$, where $Q(z) = P(T(z))$.
6. (Asymptotic geometry of Newton's map) Let $P(z)$ be a monic polynomial of degree n such that all its roots are contained in the open unit disc $\mathbb{D}$. For $|z| \geq 1$, prove the following inequalities:
   (i) $|N_P(z) - (\frac{n-1}{n})z| < \frac{1}{n}$.
   (ii) $|N_p(z)| < |z|$.
7. A generalization of Newton's method is the so-called relaxed Newton's method. Now we consider the iteration of the function $N_c(z) = z - cf(z)/f'(z)$. Determine the values of the complex constant c such that every zero of f is an attracting fixed point of N_c (consider the case of a multiple zero).
8. If $f(z) = e^z - 1$, $N(z) = z - 1 + e^{-z}$ is the corresponding Newton's map. Show that $\mathbb{D}$ is contained in the attractive basin of the root $z^* = 0$.

C Computer graphics of complex dynamics

In this appendix, we explain some basic rudiments on Matlab images that are needed to obtain the beautiful computer graphics we show throughout the book. We will employ two different types of images, RGB images and pseudoimages produced by the Matlab function **pcolor**. All graphics we show have been obtained by either of those methods.

C.1 Matlab images

(1) ***RGB images***

An $m \times n$ image has $m \cdot n$ pixels arranged in m rows and n columns. An RGB image, sometimes referred to as a truecolor image, is stored in MATLAB as an $m \times n \times 3$ data array that defines red, green, and blue color components for each individual pixel. RGB images do not use a palette. The color of each pixel is determined by the combination of the red, green, and blue intensities stored in each color plane at the pixel's location. Graphics file formats store RGB images as 24-bit images, where the red, green, and blue components are 8 bits each.

An RGB MATLAB array can be of class double, uint8, or uint16. In an RGB array of class double, each color component is a value between 0 and 1. A pixel whose color components are $[0, 0, 0]$ is displayed as black, and a pixel whose color components are $[1, 1, 1]$ is displayed as white. Next we list other examples:

- Red pixel $[b, 0, 0]$. It gets darker as b decreases.
- Green pixel $[0, b, 0]$. It gets darker as b decreases.
- Blue pixel $[0, 0, c]$. It gets darker as c decreases.
- Yellow pixel $[1, 1, 0]$.
- Orange pixel $[1, b, 0]$. It gets darker as b decreases.
- Grey pixel $[a, a, a]$. It gets darker as a decreases.
- Magenta pixel $[1, 0, 1]$. $[1, 0, c]$ gets darker as c decreases.

Given 3 $m \times n$ matrices R, G, and B, whose elements are

$$0 \le R(i,j), G(i,j), B(i,j) \le 1 \quad \text{for all } 1 \le i \le m,\ 1 \le j \le n,$$

with the Matlab code

```
I = zeros(m,n,3);
I(:,:,1) = R; I(:,:,2) = G; I(:,:,3) = B;
```

we have created an image whose name is I and such that the color of each pixel (i,j) is $[R(i,j), G(i,j), B(i,j)]$. **imshow** displays the image in the screen.

https://doi.org/10.1515/9783111689685-014

(2) *Pseudocolor plots*

A pseudocolor plot is a rectangular array of cells with colors determined by C, using each set of four points adjacent in C to define a cell.

pcolor(x,y,C) draws a pseudocolor plot of the elements of C at the locations specified by x and y. The plot is a logically rectangular, two-dimensional grid with vertices at the points (x(i,j), y(i,j)). The matrices x, y, and C have the same size.

colormap(name) sets the colormap for the current figure to the colormap specified by name. The new colormap uses the same number of colors as the current colormap and affects all axes and charts in the figure, unless you set an axes colormap separately. There are many possibilities for **name**: jet, hot, turbo, summer, winter, parula, cool, copper, etc. Furthermore, one may choose a specific number of colors. For example, **colormap**(winter(5)) uses 5 colors from the winter colormap. With shading interp, each cell is colored by a bilinear interpolation of the colors at its four vertices, using all elements of C.

To make a picture of a set S contained in the rectangular region $[a,b] \times [c,d]$, we will create a suitable grid putting

```
x = linspace(a,b,601); y = linspace(c,d,601); [x,y] = meshgrid(x,y);
```

In this way, the size of the image will be 601×601. Having chosen a maximum number of iterations N, let $C(i,j)$ be the number of iterations we need to decide that the point $(x(i,j), y(i,j))$ does not belong to S. Once we have determined the matrix C, we will use the following Matlab code:

```
hold on
colormap(name)
pcolor(x,y,C);
shading interp;
hold off
```

C.2 RGB images of filled Julia sets

To fix ideas, suppose we are interested in making a bicolor picture of the Julia set $\mathbb{J}(c)$ of the function $f_c(z) = z^4 + c$. The simplest way consists of making a graphic of the filled Julia set of f_c and visualizing $\mathbb{J}(c)$ as the boundary of the filled set. In view of Theorem 6.5.1 and the equality

$$\frac{f_c(z)}{z^4} = 1 + \frac{c}{z^4},$$

it is easy to deduce that $|f_c(z)| > |z|$ for $|z| > R(c)$, being $R(c) = \max\{2, \sqrt[4]{2|c|}\}$.

(1) *Bicolor images*

Having chosen a maximum number of iterations N, we consider the grid:

```
x = linspace(-R(c),R(c),901); y = linspace(R(c),-R(c),901);
[x,y] = meshgrid(x,y);
```

According to Theorem 6.5.2, we proceed as follows: for each z in the grid,

1. If $|f_c^k(z)| > R(c)$ for some $k \leq N$, then z does not belong to the filled Julia set and we plot a yellow point at z.
2. If $|f_c^k(z)| \leq R(c)$ for all $k \leq N$, we accept that z belongs to the filled Julia set and plot a red point at z.

The red region is an approximate plot of the filled Julia set of $f_c(z) = z^4 + c$. The next Matlab function **julia1** produces the required graphics.

Code 1

```
function [ ] = julia1(c)
% JULIA1 produces a bicolor image of the filled Julia set of
% F(z) = z^4 + c

% Step 1: Fix N and the grid, and declare the matrices R, G, and B
N = 1500; R(c) = max{2, (2|c|)^(1/4)};
x = linspace(-r,r,901); y = linspace(r,-r,901);
[x,y] = meshgrid(x,y); [m,n] = size(x);
R = zeros(m,n); G = R; B = R; % we start with black pixels

% Step 2: for each z in the grid, we calculate the iterations F^n(z)
% and determine the elements of R, G, and B
for k = 1 : m
for h = 1 : n
a = x(k,h) + y(k,h)*i
b = a^4 + c; count = 1;
while (abs(b)<=r)&(count<=N)
a = b; b = a^4 + c; count = count + 1;
end
if count <= N
R(k,h) = 1; G(k,h) = 1;
else
R(k,h) = 1;
end
end
```

```
end

% Step 3: we define the image I
I = zeros(m,n,3);
I(:,:,1) = R; I(:,:,2) = G; I(:,:,3) = B;
clear R G B x y a b
imshow(I,'border','tight')
```

You can see the filled Julia set of f_c for $c = -0.242 + 0.43i$ in Figures C.5 and C.6.

Remarks C.2.1. 1. In the first step, we have set y = linspace(R(c), -R(c), 901) and this needs some explanation. Note that [x,y] = meshgrid(x,y) produces two matrices x and y so that, when we read them from top to bottom and from left to right, the corresponding points of the grid run the square $[-R(c), R(c)] \times [-R(c), R(c)]$ from the bottom to the top. However, pixels run the image from the top to the bottom. That is why we use $y = linspace(R(c), -R(c), 901)$.

2. Obviously, with this Matlab code we obtain an approximate picture of the filled Julia set. This approximation improves as N (the maximum number of iterations) increases.

(2) *RGB images with a wider range of colors*

Introducing a wider range of colors, we get more spectacular graphics and, in fact, more information about the behavior of the orbits. Concretely, we may choose the color of a point z in the grid depending on the number of iterations we need to decide that z does not belong to the filled Julia set.

For it, one may proceed in the following way. Choose positive integers N_j such that $0 < N_1 < N_2 < \cdots < N_p = N$ and assign a different color to each interval $[N_j, N_{j+1})$. For each z in the grid, if $|f_c^k(z)| > r(c)$ for some k in $[N_i, N_{i+1})$, we plot z with the color that corresponds to $[N_j, N_{j+1})$.

The following Matlab function produces an image of this type.

Code 2

```
function [ ] = julia2(c)
% JULIA2 produces the filled Julia set of F(z) = c cos(z^4) with a wider
% range of colors. Recall that, if F is transcendental, the Julia set is
% unbounded, then we can only see its shape in some rectangle [a,b]×[c,d].
x = linspace(-2,2,901); y = linspace(2,-2,901);
[x,y] = meshgrid(x,y); N = 500;
[m,n] = size(x); R = zeros(m,n); G = R; B = R; C = R;
for k = 1:m
for h = 1:n
a = x(k,h) + y(k,h)*i
b = c* cos(a^4); count = 1;
```

```
while (abs(b) < 10)&(cont <= N)
a = b; b = c*cos(a^4); count = count + 1;
end
C(k,h) = count;
if C(k,h) <= 3
R(k,h) = 1; G(k,h) = 0; B(k,h) = 0;
elseif (C(k,h) > 3)&(C(k,h)< 7)
R(k,h) = 0; G(k,h) = 1; B(k,h) = 1;
elseif (C(k,h) > = 7)&(C(k,h) <= 20)
R(k,h) = 0; G(k,h) = 0; B(k,h) = 1;
elseif (C(k,h) >= 21)&(C(k,h) <= 25)
R(k,h) = 0; G(k,h) = 1; B(k,h) = 0;
elseif (C(k,h) > 25)&(C(k,h) <= 30)
R(k,h) = 1; G(k,h) = 0.5;
elseif (C(k,h) > 30)&(C(k,h) <= 40)
G(k,h) = 1; R(k, h) = 1;
elseif (C(k,h) >= 40)&(C(k,h) <= 500)
R(k,h) = 1; G(k,h) = 0; B(k,h) = 1;
end
end
end
I = zeros(m,n,3);
I(:,:,1) = R; I(:,:,2) = G; I(:,:,3) = B;
clear C R G B x y a b
imshow(I,'border','tight')
```

The resulting image, in the case $c = 1.25i$, appears in Figure C.8. Other examples of filled Julia sets are shown in Figures C.9, C.10, C.13, and C.14. Sometimes the behavior of orbits results in an aesthetical image. In Figures C.11, and C.12, we see two examples of this fact.

C.3 Pcolor images

Sometimes graphics obtained using **pcolor** are spectacular. The next Matlab function **juliaz4** follows this method to get a plot of the filled Julia set. We consider again the family $f_c(z) = z^4 + c$.

Code 3

```
function [ ] = Juliaz4(c)
% JULIAZ4 produces a graphic of the filled Julia set of
% F(z) = z^4 + c using pcolor
% Step 1: We fix N, the grid, and declare the matrix C
```

```
N = 500; r = max([2,(2*abs(c))^(1/4)]);
x = linspace(-r,r,901); y = linspace(-r,r,901);
[x,y] = meshgrid(x,y); [m,n] = size(x); C = zeros(m,n);
% Step 2: For each z in the grid, we calculate the iterations and
% determine C
for k = 1 : m
for h = 1 : n
z = x(k,h) + y(k,h)*i; a = z; b = z^4 + c; count = 1;
while (abs(b) < = r)&(count < = N)
a = b; b = b^4 + c; count = count + 1;
end
C(k,h)= count;
end
end
% Step 3: We create the image with pcolor
hold on
colormap(turbo) % other choices: summer,winter,jet,cool,hot,parula,etc.
pcolor(x,y,C)
shading interp
hold off
```

Figure C.7 has been obtained in this way.

C.4 The inverse iteration method

This method, based on Theorem 6.3.4, is effective when f is a polynomial with a small degree. Otherwise, the number of preimages may grow uncontrollably fast.

The next Matlab function **juliaquad4** follows this method to get a plot of $J(c)$. First of all, we have to look for some point z_0 in $J(c)$. We may often determine a repelling fixed point of Q_c. Indeed, given c, if one of the square roots of $1-4c$ does not belong to $\overline{\mathbb{D}}(-1,1)$, then $|1+\sqrt{1-4c}| > 1$ and, therefore, $z_0 = (1+\sqrt{1-4c})/2$ is a repelling fixed point of Q_c since $|Q_c'(z_0)| = 2|z_0| > 1$ (it is obvious that z_0 is not an exceptional value). Once we have found $z_0 \in \mathbb{J}(c)$, we proceed to calculate the preimages, $z = \pm\sqrt{z_0 - c}$.

Code 4

```
function [ ] = juliaquad4(c)
% JULIAQUAD4 produces the Julia set of F(z) = z^2 + c drawing
% the preimages of a point z0 ∈ J(c)
N = 20; zo = (1 + sqrt(1 - 4*c))/2; z = sqrt(zo - c); z1 = -z;
X = [real(z) real(z1)]; Y = [imag(z) imag(z1)];
for n = 2 : N
```

```
s = 0;
for q = 1 : n - 1
s = s + 2^q;
end
for j = 0 : 2^(n - 1) - 1
w = X(s - j) + Y(s - j)*i; z = sqrt(w - c); z1 = -z;
X = [X real(z) real(z1)]; Y = [Y imag(z) imag(z1)];
end
end
plot(X,Y,'.')
```

In Figures C.1, C.2, C.3, and C.4, we show images of the Julia sets of $Q_c(z) = z^2 + c$ obtained applying the inverse iterate method.

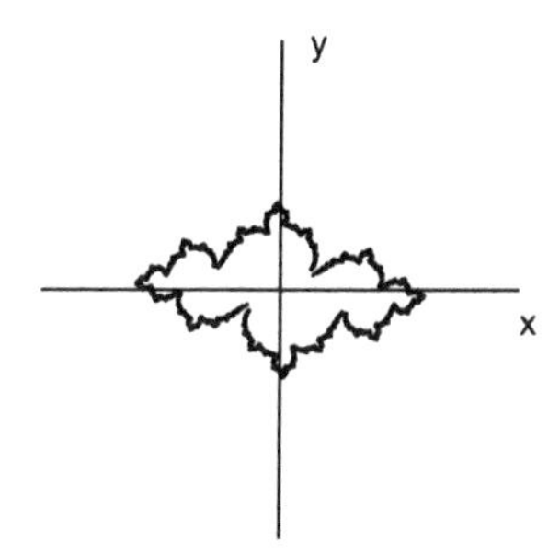

Figure C.1: Julia set of $f(z) = z^2 - 0.744336 + 0.1211i$.

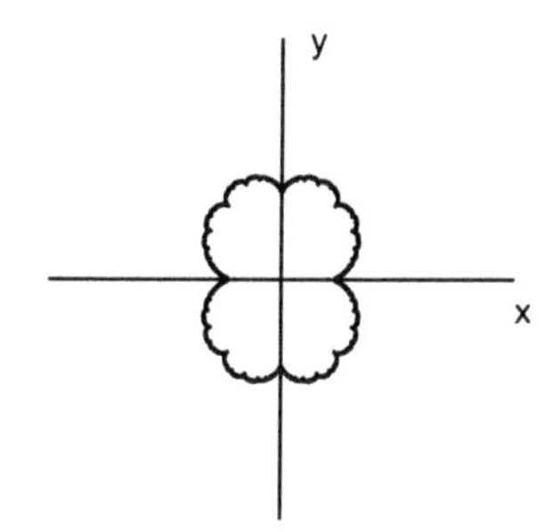

Figure C.2: Julia set of $f(z) = z^2 + 0.25$ (cauliflower).

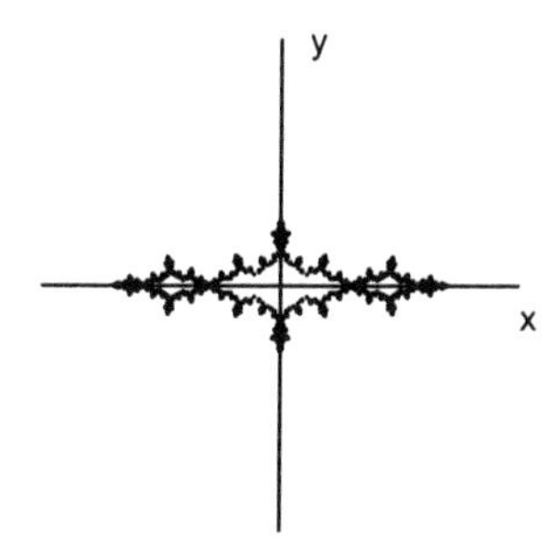

Figure C.3: The Julia set of $f(z) = z^2 - 1.3$ (airplane).

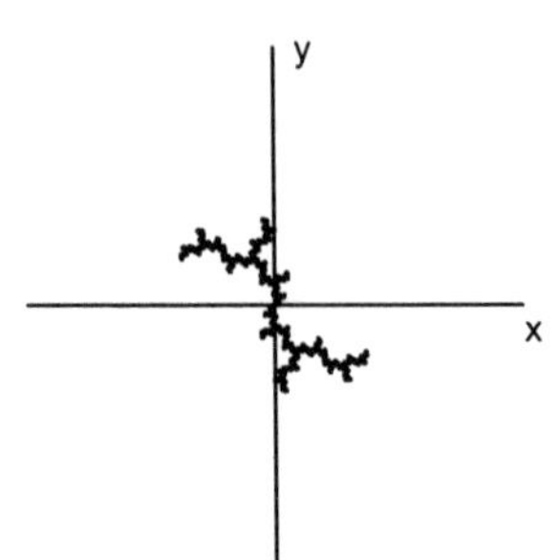

Figure C.4: The Julia set of $f(z) = z^2 + i$ (dendrite).

C.5 Gallery

In this section we show beautiful images we have obtained using the above Matlab codes.

Figure C.5: In red, the filled Julia set of $f(z) = z^4 - 0.242 + 0.43i$.

Figure C.6: An enlargement centered at $z = 0.517 + 0.4i$ in the above image.

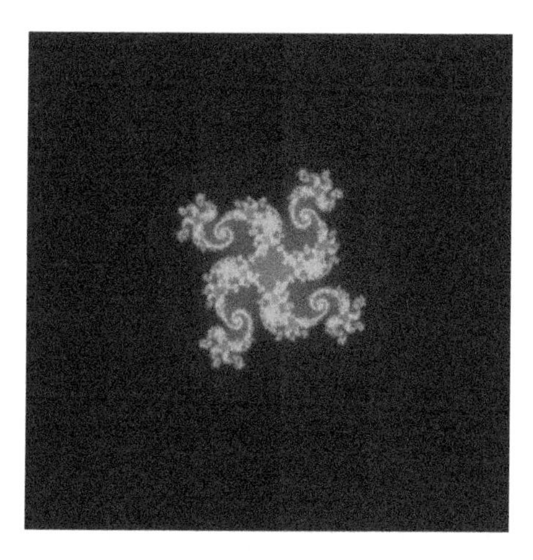

Figure C.7: The filled Julia set of $f(z) = z^4 + 0.5 + 0.65i$ (using the Matlab function **pcolor**).

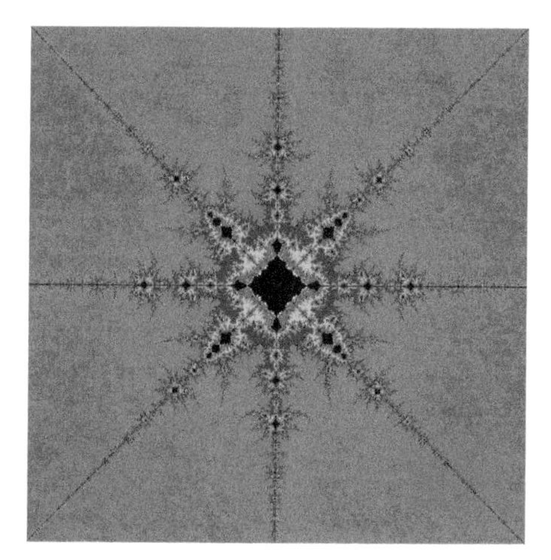

Figure C.8: The filled Julia set of $f(z) = 1.25i\cos(z^4 + P)$ for $P = 11\pi i/144$ (black).

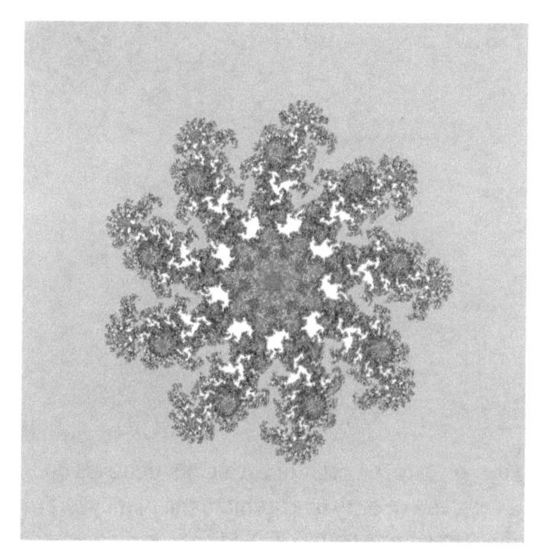

Figure C.9: The filled Julia set of $f(z) = z^9 - 0.753 + 0.04985i$ (red).

Figure C.10: The filled Julia set of $f(z) = (-2.4 - 2.2i)e^z$ (black).

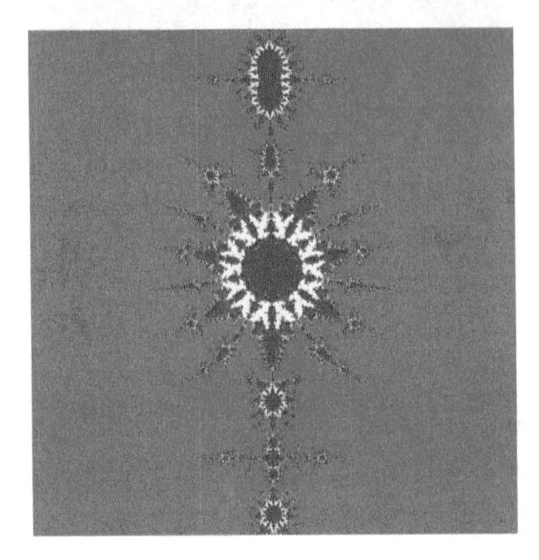

Figure C.11: The color of each point depends on the escape velocity of its orbit to the infinity; $f(z) = 2.925i\cosh^7(z + 7\pi i/6)$.

Figure C.12: The color of each point depends on the escape velocity of its orbit to the infinity; $g(z) = z^6 + 2z^4 + 2z^2 + 0.216 + 0.225i$.

Figure C.13: The filled Julia set (black) of $f(z) = 1.5\cos(z^2)$.

Figure C.14: The filled Julia set (black) of $f(z) = 1.25i\cos(z^4)$.

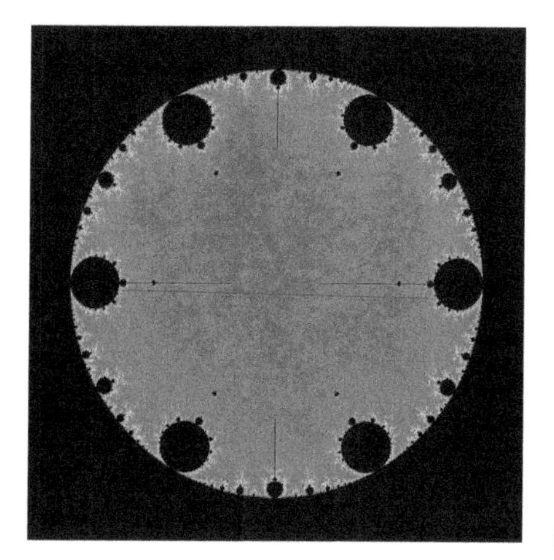

Figure C.15: Parameter plane of $f(z) = c^{-3}\sin(z)$.

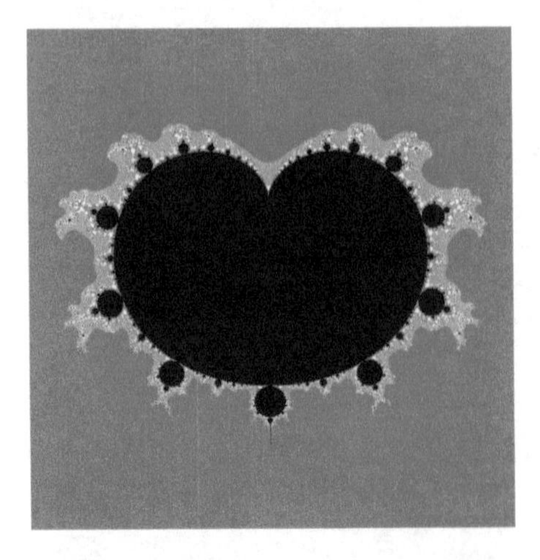

Figure C.16: Parameter plane of $f(z) = z^2 - ic/(1 - ic)$.

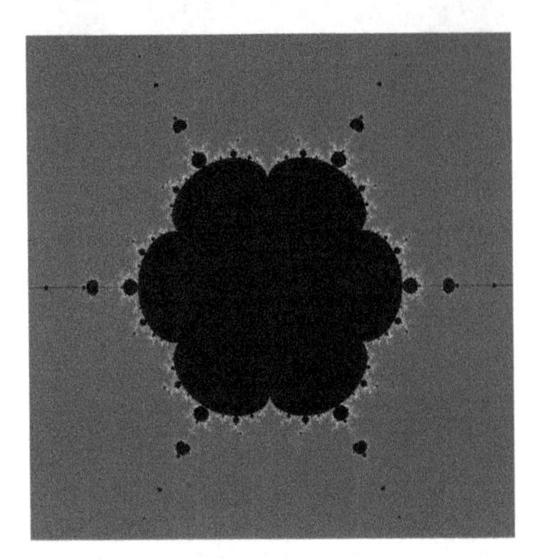

Figure C.17: Parameter plane (black) of $f(z, c) = c^3 \cos(z)$.

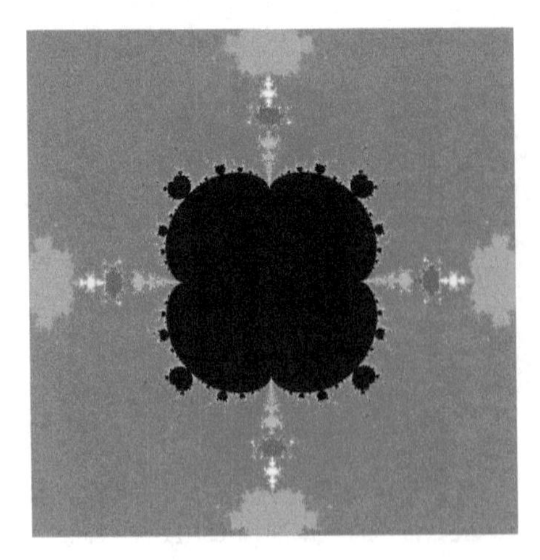

Figure C.18: Parameter plane (black) of $f(z, c) = z + c^2 + \sin z$.

Figure C.19: The regions of convergence to each root of the equation $z^8 - 1 = 0$.

Figure C.20: Convergence regions towards the roots $\pm i\,0.8597\ldots$ of $z^2 + \cos(z^2) = 0$.

D Hints to selected exercises

Chapter 1

Exercise 1. To prove the inequality $2\omega/\pi \leq \sin\omega$ $(0 \leq \omega \leq \pi/2)$, consider the function $g(\omega) = \sin\omega - (2/\pi)\omega$ for ω in $[0, \pi/2]$.

Exercise 2. To prove that h is surjective notice that, if $w = z + 1/z$, then $z^2 - zw + 1 = 0$ and, therefore, we have $z = w/2 \pm (\sqrt{w^2 - 4})/2$. Thus

$$z = \frac{w}{2} \pm \frac{\sqrt{4 - w^2}}{2} i \quad \text{and} \quad |z|^2 = \frac{w^2}{4} + \frac{(4 - w^2)}{4} = 1$$

for $w \in (-2, 2)$. This proves that, if $z \in A$, then $w = h(z)$ does not belong to the interval $(-2, 2)$. Now note that the above relations imply $z = \pm 1$ whenever $w = \pm 2$ and this proves that $h(A) \subset \mathbb{C} \setminus [-2, -2]$. Finally, if $z_1 = w/2 + \sqrt{w^2 - 4}/2$ and $z_2 = w/2 - \sqrt{w^2 - 4}/2$, then $z_1 + z_2 = w$ and $z_1 z_2 = 1$. By contradiction, it is easy to show that one of them has modulus greater than 1.

Exercise 3. Take $c = -a_1^2 + a_2 a_0 + a_1$.

Exercise 4. $\pi^{-1}(x_1, x_2, x_3) = (\frac{x_1}{1-x_3}, \frac{x_2}{1-x_3})$.

Exercise 5. $T(x_1, x_2, x_3) = \frac{x_1 + x_2 i}{1 - x_3}$ and $T^{-1}(z) = (\frac{\mathbb{R}e(z)}{1+|z|^2}, \frac{\mathbb{I}m(z)}{1+|z|^2}, \frac{|z|^2}{1+|z|^2})$.

Exercise 9. Study the conjugate $g(w)$ of f through the transformation $w = 1/z$.

Chapter 2

Exercise 1. The proof follows easily arguing by contradiction (notice that $\kappa[f](z) = \kappa[1/f](z)$).

Exercise 2. Use Cauchy's estimate for $|f'(z)|$ and Marty's theorem.

Exercise 4. First, using De Moivre's formula, show that $(1 + z/N)^N \to e^z$ pointwise on $\mathbb{C}$. Next, prove that $(1 + z/N)^N$ is uniformly bounded on every compact subset of $\mathbb{C}$ and apply Exercise 2.

Exercise 5. (ii) Show that $\mathbb{F}$ is normal. (iii) Apply Schwarz's lemma to f and f/z, successively.

Chapter 3

Exercise 1. Let $a_1, \ldots, a_4$ be distinct and such that $E_1(a_k) = E_2(a_k)$ for $k = 1, \ldots, 4$. Also assume that f_1, f_2 are neither both constant nor identical. Applying (C.15) with $q = 4$, we obtain $(3 + o(1))T(r, f_j) \leq \sum_{k=1}^{4} N_k(r)$, where $N_k(r) = \overline{N}(r, 1/(f_j(z) - a_k))$ for $k = 1, \ldots, 4$ and $j = 1, 2$. On the other hand, we have

https://doi.org/10.1515/9783111689685-015

$$T(r,(f_1-f_2)^{-1}) \le T(r,f_1)+T(r,f_2)+O(1) \le \frac{2}{3+o(1)}\sum_{k=1}^{4} N_k(r)+O(1).$$

Finally, notice that $\sum_k N_k(r) \le \overline{N}(r,(f_1-f_2)^{-1})$ since each common root of $f_j(z) = a$ for $j = 1,2$ is a pole of $(f_1-f_2)^{-1}$. Combining all these facts, we deduce

$$\left(1-\frac{2}{3+o(1)}\right)\sum_{k=1}^{4} N_k(r) \le O(1),$$

which is a contradiction because f_1, f_2 are not constant.

Exercise 3. (ii) Notice that, if z_0 is a root of $e^z = a$, then all other roots are of the form $z_0 + 2\pi mi$, with m being an integer. To simplify, assume that a is a real number.

Exercise 4. Notice that $\overline{N}(r,a) \le (1/m)N(r,a) \le (1/m)T(r,f)+O(1)$. Therefore, $\Theta(a) \ge 1-1/m$. Finally, use Nevanlinna's inequality.

Exercise 5. If z_0 is a pole of f of order p, then $f^{k)}$ has a pole of order $p+k \ge k+1$ at z_0. Therefore, $\overline{N}(r,f^{k)}) \le (1/(k+1))N(r,f^{k)}) \le T(r,f^{k)})$. This implies that $\Theta(\infty,f^{k)}) \ge k/(k+1)$.

Chapter 4

Exercise 2. Consider $r = \log(1/|\lambda|)$. Notice that $z = \log_p(1/\lambda)$ is a repelling fixed point of f_λ and, consequently, r cannot be improved for every λ with a positive real part.

Exercise 3. Apply the implicit function theorem.

Exercise 7. It suffices to consider the case of an attracting fixed point z^*. By contradiction, consider a Jordan curve γ in the immediate basin $\Omega_0(f,z^*)$ such that there is some point $z_0 \in \operatorname{int}(\gamma) \cap \Omega_0(f,z^*)^c$. Now apply the Cauchy integral formula to each f^n.

Exercise 10. (i) Prove that $P(z)$ is conjugate to $Q(z)$ through a certain affine map $\phi(z) = Az+B$.

Chapter 5

Exercise 2. If $h(\zeta_1) = h(\zeta_2)$, then $f(h(\zeta_1)) = f(h(\zeta_2))$ and, therefore, $h(\lambda\zeta_1) = h(\lambda\zeta_2)$. Repeating the argument, we obtain $h(\lambda^n\zeta_1) = h(\lambda^n\zeta_2)$ for $n \in \mathbb{N}$. As $\{\lambda^n : n \in \mathbb{N}\}$ is dense in the circle $|z| = 1$, we have $h(z\zeta_1) = h(z\zeta_2)$ for all $|z| = 1$. Finally, consider the function $g(z) = h(z\zeta_1) - h(z\zeta_2)$ defined on a suitable disc $\mathbb{D}_\varepsilon$.

Exercise 3. Apply the real mean value theorem to f.

Exercise 4. Consider the homeomorphism $T : x \in \mathbb{R}/\mathbb{Z} \to e^{2\pi xi} \in S^1$. Given an integer $N > 1$, cut the circle into N half-open arcs of length $1/N$. Notice that at least two of the $Q+1$ numbers $T(0), T(x), T(2x), \ldots, T(Nx)$ must belong to the same arc.

Exercise 5. Given $0 < \varepsilon < 1$, choose $q \in \mathbb{N}$ such that $|\lambda^q - 1|^{1/d^q} < \varepsilon$. To simplify, we consider a monic polynomial $f(z) = z^d + \cdots + \lambda z$. It is easy to show that $f^q(z) = z^{d^q} + \cdots + \lambda^q z$. The fixed points of f^q are the roots of $z^{d^q} + \cdots + (\lambda^q - 1)z = 0$. The product of the nonzero roots is equal to $\pm(\lambda^q - 1)$. One of these roots z verifies $0 < |z|^{d^q - 1}| \leq |\lambda^q - 1|$.

Exercise 6. (i) Notice that θ_i is a rational number with denominator $p < q$. Then the procedure must terminate after at most q steps.

(ii) Use induction and the equality

$$[a_1, a_2, \ldots, a_n] = [a_1, a_2, \ldots, a_{n-2}, a_{n-1} + (1/a_n)].$$

Exercise 8. Find the rational numbers closest to π with denominators 1, 2, or 3.

Chapter 6

Exercise 1. (ii) Show that $\mathbb{F}(g) = \mathbb{F}(f)$.

Exercise 6. Prove the implication: z is repelling k-periodic for $f \Rightarrow e^z$ is repelling k-periodic for g.

Exercise 7. Find the repelling fixed points of f.

Exercise 8. If $R(z) = a_{n-1}/z + \cdots + a_1/z^{n-1} + a_0/z^n$, prove that $|R(z)| \leq (\sum_0^{n-1} |a_k|)/R$ for all $|z| > R$.

Exercise 10. Given $w \in \mathbb{C} \setminus [-1, 1]$, choose $z = x + iy$ such that $w = \cos z$ (note that $y \neq 0$). Show and then use the inequality $|\cos(x + iy)|^2 = |\cos^2 x \cosh^2 y + i \sin^2 x \sinh^2 y|^2 \geq \sinh^2 y$.

Exercise 11. Use Exercise 1.

Exercise 12. Show that $P(\mathbb{D}) = \mathbb{D}$ and then apply the minimum modulus principle to $P/T_1 \cdot T_2 \cdot T_k$, where each T_j is a Möbius transformation from $\mathbb{D}$ to itself and the zeros of $T_1 \cdot T_2 \cdot T_k$ coincide with those of P in $\mathbb{D}$.

Chapter 7

Exercise 1. As in Example 7.6.2, the proof is based on Proposition 7.6.1. Consider the function $g(z) = z + 2h(e^z)^2$ and notice that $g(z + 2\pi i) = g(z) + 2\pi i$.

Exercise 3. As in Example 7.5.2, take $V = \mathbb{D}_r$ contained in the basin of attraction $\Omega(g, 0)$, with $g(w) = cwe^w$, and consider the half-plane $H = \{z \in \mathbb{C} : \mathbb{R}e(z) < \log r\}$. Notice that $H \subset \pi^{-1}(V)$, where $\pi(z) = e^z$.

Exercise 4. Take $g(w) = w^2 e^w$ and apply the logarithmic change of variable $w = e^z$.

Exercise 5. Without loss of generality, we may assume that $\mathbb{F}(f)$ has a Siegel disk U. As f is injective in U, Picard's theorem asserts that for all $z \in U$, except at most for one element, the equation $f(\zeta) = z$ has infinitely many solutions and only one of them may belong to U. Obviously, those solutions are in $\mathbb{F}(f)$. So there exists a Fatou component V containing a point ζ such that $f(\zeta) = z \in U$. This V is preperiodic.

Exercise 6. Set $g(z) = z + (\lambda - 1)(e^z - 1)$ and note that the complex numbers $z_n = 2\pi ni$ $(n \in \mathbb{Z})$ are fixed points of g with multiplier λ. Then each z_n is the fixed point in a Siegel disc S_n for g. By Proposition 7.6.1, $\mathbb{F}(f) = \mathbb{F}(g)$ and, therefore, the S_n's are components of $\mathbb{F}(f)$. As $f(2\pi ni) = 2\pi(n+1)i$, it follows that S_1 is a wandering domain for f so that f is univalent in S_1 (Theorem 7.4.2 tells us that g is univalent in S_1).

Exercise 7. Apply the maximum modulus principle.

Exercise 9. Apply Theorems 6.5.4 and 6.5.5.

Exercise 10. For a fixed $a \in U$, let b be a solution of the equation $P(z) = a$. Note that $b \notin \Omega_0(P, z^*) \cup U \cup \Omega(P, \infty)$. Then b belongs to another Fatou component U_1. Since $a = P(b)$, it follows that $P(U_1) \subset U$. Inductively, this argument allows us to obtain an infinite sequence of Fatou components (U_n).

Chapter 8

Exercise 1. Take a suitable curve $\gamma(t)$ tending to ∞ and use the fact that the integral $\int_0^\infty \frac{\sin r}{r}\, dr$ is convergent.

Exercise 4. To simplify, we consider the case $f(z) = \cos z$. Suppose there is a curve $\gamma(t) = x(t) + iy(t)$ satisfying $\lim_{t\to\infty} \gamma(t) = \infty$ and $\lim_{t\to\infty} f \circ \gamma(t) = a_1 + ia_2$, then

$$(e^{-y(t)} + e^{y(t)}) \cos x(t) \to 2a_1 \quad \text{and} \quad (e^{-y(t)} - e^{y(t)}) \sin x(t) \to 2a_2.$$

In the case $y(t) \to \infty$, the first equality yields $\cos x(t) \to 0$ and, in view of the second, we have a contradiction. Then we may assume that $y(t)$ is bounded and, in consequence, $x(t) \to \infty$. Finally, notice that $(e^{-y(t)} + e^{y(t)}) \cos x(t) > \cos x(t)$ for $\cos x(t) > 0$ and $(e^{-y(t)} + e^{y(t)}) \cos x(t) < \cos x(t)$ for $\cos x(t) < 0$. Now proceeding as in Example 8.1.2(3), we may conclude that $\lim_{t\to\infty} (e^{-y(t)} + e^{y(t)}) \cos x(t)$ does not exist.

Exercise 6. Assume that $\lambda ne < 1$.

Exercise 9. If there is a neighborhood U of z^* such that $U \cap S^+(f) = \emptyset$, then there exists an analytic branch h_n of each $(f^n)^{-1}$ defined on U such that $h_n(z^*) = z^*$. Now apply Lemmas 8.1.6 and 8.1.7.

Exercise 10. (ii) Show that f has no critical points outside the strip $\{z = x + iy : -r < y < r\}$, and its possible critical values in that strip are bounded.

Exercise 11. It suffices to consider the case of a fixed component. Using that U is bounded and g^k is uniformly continuous on every compact set, show that (f^n) is normal on each region $g^k(U)$.

Exercise 12. Show that $S^+(f)$ is contained in the immediate basin of attraction of the origin.

Chapter 9

Exercise 1. Suppose D is a disc with radius $r > 0$ contained in $\mathbb{J}(E_\lambda)$ and show that $\mathbb{R}e(E_\lambda^n(z)) \geq p_\lambda$ for all $z \in D$ and $n \in \mathbb{N}$.

Exercise 2. For $z = x + iy \in S$, we have $e^x \cos y \geq e^x/2 \geq x + 1 - \log 2$.

Exercise 3. If $z \in S$ is complex such that $\mathbb{R}e(z) > 1$, then

$$|e^x \sin y| \geq e^x\left(\frac{2|y|}{\pi}\right) > c|y|,$$

for some $c > 1$. Thus $|\mathbb{I}m(E^n(z))|$ must grow as n increases. Having in mind the previous exercise, it is easy to conclude the proof.

Exercise 4. To simplify, we consider the case $k = 1$. If $z = x + iy$ is a repelling fixed point in S_k, then $|e^z| = e^x > 1$. This shows that x must be positive. On the other hand, it follows from the equality $e^z = z$ that

$$y = \sqrt{e^{2x} - x^2}, \quad y = \arccos(xe^{-x}) \quad \text{and} \quad y \in (2\pi, 2\pi + \pi/2).$$

Now it suffices to prove that those curves intersect.

Exercise 6. Put $S_\lambda(iy) = ig_\lambda(y)$ for $y \in \mathbb{R}$, where $g_\lambda(y) = \lambda \sinh y$, and find the fixed points of g_λ.

Exercise 7. (iii) Show that $\varepsilon > 0$ may be chosen such that $|S_\lambda'(z)| \geq \mu > 1$ for all $z \in B(N)$.

Exercise 8. Apply Theorems 6.2.3 and 8.3.4.

Exercise 9. Prove that $f \in \mathcal{S}$ and all singular values are in $\mathbb{D}$.

Chapter 10

Exercise 1. $C_1 = \mathbb{D} \cup \{e^{-z} : |z+1| < 1\}$. If $\lambda = e^{-z}$ with $|z+1| < 1$, z is an attracting fixed point for F_λ, but the attracting fixed point is the origin for $|\lambda| < 1$.

Exercise 5. Consider the equation $f(z, c) = Q_c^n(z) - z = 0$ and apply the implicit function theorem at (z_0, c_0).

Exercise 6. By contradiction, show that (g_n) is uniformly convergent to ∞ on compact subsets of the complement of $\mathbb{M}$.

Exercise 7. Find an affine transformation T such that P_μ is conjugate to Q_c through T and then apply Theorem 10.3.5.

Exercise 9. The fixed points of Q_c are fixed for Q_c^2.

Chapter 11

Exercise 3. Use the explicit formula of the escape radius.

Exercise 4. (iii) As an example, consider $P_\mu(z) = \mu z(1 - z)$ and $\mu_n = 3n/(n + 1)$.

Exercise 5. Let $O(z_N)$ be the corresponding attracting k-periodic orbit for $P_{N,\lambda}$. Note that $P_{N,\lambda}(z) = (P_{N,\lambda})'(z)(1 + z/N)$. Then

$$1 > |(P_{N,\lambda}^k)'(z_N)| = \left|\prod_{\hat{z}\in O(z_N)} (P_{N,\lambda})'(\hat{z})\right| = \left|\frac{z_N}{1 + z_N/N}\right| \prod_{\hat{z}\in O(z_N),\hat{z}\neq z_N} \left|\frac{\hat{z}}{1 + \hat{z}/N}\right|.$$

The above relation may be expressed as follows:

$$\left|\frac{z_N}{1 + z_N/N}\right| < \prod_{\hat{z}\in O(z_N),\hat{z}\neq z_N} \left|\frac{1 + \hat{z}/N}{\hat{z}}\right| = \prod_{\hat{z}\in O(z_N),\hat{z}\neq z_N} \left|\frac{1}{N} + \frac{1}{\hat{z}}\right|.$$

Now it is easy to prove that, for each $N \in P$, there is a point of modulus less than 2 in the orbit $O(z_N)$. These points z_N may be chosen such that they have an accumulation point z_0. Prove that z_0 is a neutral k-periodic point for E_λ.

Appendix A

Exercise 1. Show that f maps $\mathbb{D}$ into the set $\mathbb{C} \setminus (-\infty, -1/4]$.

Exercise 2. The upper estimate follows applying the Koebe one-quarter theorem to f^{-1}.

Exercise 4. By Theorem 1.6.1, the Möbius transformation $T(w) = \frac{z+w}{1+\bar{z}w}$ is an automorphisms of $\mathbb{D}$.

Appendix B

Exercise 3. If $|z - 1| < r$, show that

$$|N_f(z) - 1| \leq |z - 1|\frac{3r + 2r^2}{3(1 - r)^2}.$$

We must take $r \in (0, 1)$ so that $k = (3r + 2r^2)/(3(1 - r)^2) < 1$.

Exercise 6. If $P(z) = \prod_{k=1}^n (z - \zeta_k)$, notice that each $z - \zeta_k$ belongs to $\mathbb{D}(z)$, and the set $D = \{w : 1/w \in \mathbb{D}(z)\}$ is a certain disc, say $\mathbb{D}(z_0, r)$. Then $\sum_k 1/(z - \zeta_k) \in \mathbb{D}(nz_0, nr)$. Taking the inverse, again we deduce that $1/(\sum_k 1/(z - \zeta_k)) \in \mathbb{D}_{1/n}(z/n)$.

Bibliography

[1] Aarts J., Oversteegen L. The Geometry of Julia sets. Trans. Am. Math. Soc. 338, 897–918 (1993).
[2] Ahlfors L. Complex Analysis. McGraw-Hill Co. (1978).
[3] Alexander D. S. A History of Complex Dynamics. Springer Fachmedien Wiesbaden (1994).
[4] Baker I. N. The existence of fixpoints of entire functions. Math. Z. 73, 280–284 (1960).
[5] Baker I. N. Repulsive fixpoints of entire functions. Math. Z. 104, 252–256 (1968).
[6] Baker I. N. Limits functions and sets of non-normality in iteration theory. Ann. Acad. Sci. Fenn., Ser. A 1 Math. 467, 1–11 (1970).
[7] Baker I. N. Completely Invariant Domains of Entire Functions. Math. Essays Dedicated to A. J. Macintyre (1975). Ohio Univ. Press, Athens, Ohio (1970).
[8] Baker I. N. The domains of normality of an entire function. Ann. Acad. Sci. Fenn., Ser. A 1 Math. 1, 277–283 (1975).
[9] Baker I. N. An entire function which has wandering domains. J. Aust. Math. Soc. A 22, 173–176 (1976).
[10] Baker I. N. Wandering domains in the iteration of entire functions. Proc. Lond. Math. Soc. 49, 563–576 (1984).
[11] Baker I. N., Kotus J., Lü Y. Iterates of meromorphic functions I. Ergod. Theory Dyn. Syst. 11, 241–248 (1991).
[12] Banhs J., Brooks J., Cirns G., Davis G., Stacey P. On Devaney's definition of chaos. Am. Math. Mon. 99, 332–334 (1992).
[13] Beardon A. F. Iteration of Rational Functions. Graduate Texts in Mathematics. Springer (1991).
[14] Bergweiler W., Rhode S. T. Omitted values in domains of normality. Proc. Am. Math. Soc. 123, 1857–1858 (1995).
[15] Bergweiler W. On the Julia set of analytic self-maps of the punctured plane. Analysis 15, 251–256 (1995).
[16] Bötcher L. E. The principal laws of convergence of iterates and their application to analysis. Izv. Kazan. Fiz.-Mat. Obshch. 14, 143–206 (1904) (Russian).
[17] Bodelón C., Devaney R. L., Hayes M., Roberts G., Goldberg L. R., Hubbard J. H. Hairs for the complex exponential family. Int. J. Bifurc. Chaos 09(08), 1517–1534 (1999).
[18] Branner B., Hubbard J. H. The iteration of cubic polynomial, Part I: The global topology of parameter space. Acta Math. 160, 143–206 (1988).
[19] Brooks R., Matelski P. The dynamics of 2-generator subgroups of PSL$(2, \mathbb{C})$. In: Riemann Surfaces and Related Topics, Proceedings. Stony Brook Conference, pp. 65–71 (1978).
[20] Carleson L., Gamelin T. W. Complex Dynamics. Springer (1995).
[21] Cayley A. Applications of the Newton–Fourier method to an imaginary root of an equation. Q. J. Pure Appl. Math. 16, 179–185 (1879).
[22] Cayley A. The Newton–Fourier imaginary problem. Am. J. Math. 2, 97 (1879).
[23] Cremer H. Über die Schrödersche Funktionalgleichung und das Schwartzsche Eckenabbildungsproblem. Ber. Verh. Sächs. Akad. Wiss. Leipz., Math.-Phys. Kl. 84, 291–324 (1932).
[24] Devaney R. L. Julia sets and bifurcation diagrams for exponential maps. Bull. Am. Math. Soc. 11, 167–171 (1984).
[25] Devaney R. L. Complex Dynamical System: The Mathematics Behind the Mandelbrot and the Julia Set. American Mathematical Society (1989).
[26] Devaney R. L. Exotic topology in complex dynamics. Indag. Math. 27, 1116–1126 (2016).
[27] Devaney R., Complex L. Dynamics and entire functions. In: Complex Dynamics Systems. Proceedings of Symposia in App. Math., vol. 49. Amer. Math. Soc., Providence, Rhode Island (1994).
[28] Devaney R. L. An Introduction to Chaotic Dynamical Systems. Addison-Wesley (1994).
[29] Devaney R. L., Goldberg L. R., Hubbard J. Dynamical Approximation to the Exponential Map by Polynomials. Mathematical Sciences Research Institute, Berkeley (1986).

https://doi.org/10.1515/9783111689685-016

[30] Devaney R. L., Keen L. Dynamics of meromorphic maps with polynomial Schwarzian derivative. Ann. Sci. Éc. Norm. Supér. 22, 55–81 (1989).
[31] Devaney R. L., Krych M. Dynamics of exp(z). Ergod. Theory Dyn. Syst. 4, 35–52 (1984).
[32] Devaney R. L., Targenman F. Dynamics of entire functions near the essential singularity. Ergod. Theory Dyn. Syst. 6, 489–503 (1986).
[33] Douady A. Does a Julia set depend continuously on the polynomial? In: Complex Dynamics Systems. Proceedings of Symposia in App. Math., vol. 49. Amer. Math. Soc., Providence, Rhode Island (1994).
[34] Douady A., Hubbard J. Itération des polynômes quadratiques complexes. C. R. Acad. Sci. Paris I 29, 123–126 (1982).
[35] Douady A., Hubbard J. On the dynamics of polynomial-like mappings. Ann. Sci. Éc. Norm. Supér., Paris 18, 287–343 (1985).
[36] Eremenko A. E. On the iteration of entire functions. In: Dynamical System and Ergodic Theory. Banach Center Publications, vol. 23. PWN, Warsaw, Poland (1989).
[37] Eremenko A. E., Lyubich M. Y. Iterates of entire functions. Sov. Math. Dokl. 30(3), 592–594 (1984).
[38] Eremenko A. E., Lyubich M. Y. Examples of entire functions with pathological dynamics. J. Lond. Math. Soc. 36, 458–468 (1987).
[39] Eremenko A. E., Lyubich M. Y. Dynamical properties of some classes of entire functions. Ann. Inst. Fourier 42, 989–1020 (1992).
[40] Fatou P.: Sur les equations functionelles. Bull. Soc. Math. Fr. 47, 161–271 (1919); 48, 33–94 and 208–314 (1920).
[41] Fatou P. Sur l'iteration des fonctions transcendantes entiéres. Acta Math. 47, 337–370 (1926).
[42] Hayman W. K. Meromorphic Functions. Oxford Mathematical Monographs (1966).
[43] Henrikson J. Completeness and total boundedness of the Hausdorff metric. MIT Undergrad. J. Math. (1999).
[44] Herring M. Mapping properties of Fatou components. Ann. Acad. Sci. Fenn., Math. 23, 263–274 (1998).
[45] Hille E. Analytic Function Theory II. Chelsea Publ. Co., New York (1962).
[46] Holland A. S. B. Introduction to the Theory of Entire Functions. Academic Press, New York and London (1973).
[47] Hua X. H., Yang C. Dynamics of Transcendental Functions. Asian Math. Series. Gordon and Breach Science Publishers (1998).
[48] Hurwitz A. Sur les points critiques des fonctions inverses. C. R. Séances Acad. Sci., Paris 143, 877–879 (1906).
[49] Hutchinson J. E. Fractals and self-similarity. Indiana Univ. Math. J. 30, 713–747 (1991).
[50] Iversen F. Recherche sur les fonctions inverses des fonctions meromorphes. Doctoral thesis, Helsingfors, 1914.
[51] Julia G. Memoire sur l'iteration des fonctions rationelles. J. Math. Pures Appl. 4, 47–245 (1918).
[52] Kisaka M. Local uniform convergence and convergence of Julia sets. Nonlinearity 8, 273–281 (1995).
[53] Kisaka M. On the connectivity of Julia sets of transcendental entire functions. Ergod. Theory Dyn. Syst. 18, 189–205 (1998).
[54] Koenigs G. Recherches sur les integrales de certains équations functionelles. Ann. Sci. Éc. Norm. Supér. 3(1), 1–41 (1884).
[55] Kotu J., Urbanski M. Meromorphic Dynamics. Cambridge Universiy Press (2023).
[56] Krauskopf B. Convergence of Julia sets in the approximation of λe^z by $\lambda(1 + z/d)^d$. Int. J. Bifurc. Chaos 3, 257–270 (1993).
[57] Lavaurs P. Systèmes dynamiques holomorphes: Explosion de points périodiques paraboliques. Thesis, Univ. Paris-Sud, Orsay (1989).
[58] Liao L., Yang C. C. On the Julia set of two permutable entire functions. Rocky Mt. J. Math. 35(5), 1657–1674 (2005).
[59] Mandelbrot B. The Fractal Geometry of Nature. W. H. Freeman (1982).
[60] Markushevich A. Theory of Functions of a Complex Variable, I and II. Prentice-Hall (1965).

[61] Milnor J. Dynamics in One Complex Variable. Introductory Lectures. Vieweg (2000).
[62] Misiurewicz M. On iterates of e^z. Ergod. Theory Dyn. Syst. 1, 103–106 (1981).
[63] Morosawa S., Nishimura Y., Taniguchi M., Ueda T. Holomorphic Dynamics. Cambridge University Press (2000).
[64] Morse M., Hedlung G. Symbolic dynamics. Am. J. Math. 60, 815–866 (1938).
[65] Nevanlinna R. Le thoérème de Picard-Borel et la théorie des fonctions méromorphes. Gauthiers-Villars, Paris (1929).
[66] Nevanlinna R. Analytic Functions. Springer, New York (1970).
[67] Newman M. Elements of the Topology of Plane Sets. Cambridge University Press (1939).
[68] Peitgen H. O., Richter P. H. The Beauty of Fractals: Images of Complex Dynamical Systems. Springer, Berlin (1986).
[69] Poon K. K., Yang C. C. On the Fatou components of two permutable transcendental entire functions. J. Math. Anal. Appl. 278(2), 512–526 (2003).
[70] Poon K. K., Yang C. C. Dynamical behaviour of two permutable entire functions. Ann. Pol. Math. 68, 159–163 (1998).
[71] Schröder E. Ueber unendlich viele Algorithmen zur Auflosung der Gleichungen. Math. Ann. 2, 317–365 (1870).
[72] Schröder E. Ueber iterite Functionen. Math. Ann. 3, 296–322 (1871).
[73] Schwick W. Repelling periodic points in the Julia set. Bull. Lond. Math. Soc. 29, 314–316 (1997).
[74] Siegel C. Iteration of analytical functions. Ann. Math. 43, 607–612 (1942).
[75] Steinmetz N. Rational Iteration: Complex Analytic Dynamical Systems. De Gruyter (1993).
[76] Sullivan D. Quasiconformal homeomorphisms and dynamics I. Solution of the Fatou–Julia problem on wandering domains. Ann. Math. 122, 401–418 (1985).
[77] Szusz P., Rochett A. Continued Fractions. World Scientific, (1992).
[78] Tanaguchi M. Logarithmic lifts applied to the family $\lambda z \exp(z)$. Proc. Math. Anal. Inst., Kyoto Univ. 988, 21–28 (1997).
[79] Zalcman L. A heuristic principle in complex function theory. Am. Math. Mon. 82, 813–817 (1975).

Index

https://doi.org/10.1515/9783111689685-017

www.ingramcontent.com/pod-product-compliance
Lightning Source LLC
Jackson TN
JSHW051137200325
81115JS00004B/12

* 9 7 8 3 1 1 1 6 8 9 4 3 2 *